Detroit-Mich
March. 30th 1924

That the students of steam engineering may compare present-day instructions with that of the old school. this book is presented to Dean M. E. Cooley of the University of Ann Arbor Mich. by L. H. McCracken of "M. E. Cooley Association # 7. National Association of Stationary Engineers. Detroit-Mich

LESSONS AND PRACTICAL NOTES

ON

STEAM,

THE STEAM ENGINE, PROPELLERS,

ETC., ETC.,

FOR

Young Engineers, Students, and Others.

BY THE LATE

W. H. KING, U. S. N.

REVISED BY

CHIEF ENGINEER J. W. KING, U. S. N.

[THIRTEENTH EDITION, ENLARGED.]

New York:
D. VAN NOSTRAND, Publisher,
23 MURRAY STREET & 27 WARREN STREET.

1870.

CONTENTS.

CHAPTER IV.

CAUSALTIES, ETC.

CHAPTER V.

MISCELLANEOUS.

CHAPTER VI.

WESTERN RIVER BOAT ENGINE.

CHAPTER VII.

BOILERS, ETC.

APPENDIX.

MATERIALS.

THE ELEMENTS OF MACHINERY.

INTRODUCTION.

Writing a book and then apologizing for having written it, is hardly in accordance with our convictions; but considering, nevertheless, the eminent talent which has preceded us upon the subject we have taken up, a few remarks of explanation may not be out of place. Books heretofore appearing on the steam engine, have been of two classes, or the work itself has been divided into two parts—the one for the theorist, the other for the practical man. In the one case long mathematical formulas have been produced, and in the other nothing but simple rules. The practical man, therefore, who has not had the advantage of a mathematical education, has nothing presented to him but the bare rules, which he is compelled wholly to reject, or take entirely upon trust. Besides, these works extend over numerous volumes, the study of which involve much time, labor, and expense, and which usually disheartens the practical man before he has made much progress. Having had many of these difficulties to surmount in our earlier studies of the steam engine, we were led to the course of keeping a Steam Journal, in which we noted, from time to time, as we progressed, whatever we thought important, and was made clear to our mind; and this course we would also recommend the young student; for, however well it may be to study books containing other mens' thoughts, when we write we are led to the habit of thinking for ourselves, which is of the highest importance; and, by keeping a journal, we have also the very great advantage of having always at our command, in a condensed form, those things which are the more important, and which can be referred to at any time.

Much of the present work has been taken from the Author's Journal, and the remainder has been supplied, from time to time, as he found leisure from his hours of business.

Our object has not been so much to supply wanting information, as to direct the student into the habit of thinking and reasoning for himself on those subjects which may be presented for his consideration, and which, in order that he may become eminent in his profession, he must thoroughly understand. It is not sufficient to assert that Newton said this, or somebody else said that. The reasons why they said it, and the fundamental principles upon which they based their conclusions, are necessary to be understood, in order to have a clear understanding of the subject; and if we have succeeded in making any thing more clear, or in rendering any service to that class of persons who are eagerly seeking for information, but who require some assistance to direct them in the proper channel, our only object in launching this, our little bark, on the troubled sea of authorship, is fully accomplished, conscious all the while, however, of the many imperfections it contains.

LESSONS AND PRACTICAL NOTES.

CHAPTER I.

STEAM.

STEAM is a thin, elastic, invisible fluid, generated by the application of heat to any liquid, usually water. That, however, which is generated while the water is in a state of ebullition, is alone generally termed steam, while that which is formed while the surface of the water is quiescent, is denominated vapor—a distinction, to our mind, without much difference.

The mean pressure of the atmosphere at the surface of the ocean is equal to 14.7 pounds per square inch, or is equivalent in pressure to a column of mercury 29.9212 inches in height. Under this pressure, fresh water boils at a temperature of 212° Fahrenheit.

The 212° is, however, not the total number of degrees in the steam, but simply that which is indicated by the thermometer, and which is termed sensible heat; for we all know that to raise water from the freezing to the boiling point requires a certain time, and a certain amount of fuel; and we know further, that when the water commences to boil, it does not all evaporate at once, but that the evaporation goes on gradually, and the time, and hence the fuel required to evaporate it, is much greater than that required to raise it from the freezing to the boiling point. This extra heat must have gone off somewhere, and must be in the steam, but as it is not indicated by the thermometer, it is termed *latent* heat. When the steam is reconverted into water, the latent heat becomes again sensible, which is evidenced by the large amount of water required to condense a small amount in the shape of steam. The precise ratio the one bears to the other shows the latent, compared with the sensible heat.

The subject of latent heat has been one of unusual interest, ever since the invention of the steam engine, and numerous theories have been advanced, and numerous experiments made—some of them not very carefully—in order to determine the exact law it followed; but none, up to Regnault's time, seem to have settled the subject satisfactorily. Some maintained that the latent heat of steam was a constant quantity, some that the sum of sensible and latent heat was a constant quantity, and that quantity was 1202° Fahrenheit. This was the most popular theory, and was the one generally adopted by engineers. Others, again, maintained that neither the sensible, latent, nor sum of the sensible and latent heats, were a constant quantity, but that they all varied. The exact ratio, however, in which they varied was not established until Regnault undertook his able series of experiments at the instigation of the French Government. These are the latest and most reliable experiments, and we subjoin, therefore, a table compiled from his labors, which we earnestly recommend to the attention of the reader.

REGNAULT'S EXPERIMENTS.

Degrees of heat contained in saturated steam, in Fahrenheit degrees of heat and English inches.

Temperature of the Saturated steam. (Vapor at the point of condensation.)	Corresponding elastic force		Total heat, latent heat plus sensible heat above 0° Fahrenheit.	Temperature of the Saturated steam. (Vapor at the point of condensation.)	Corresponding elastic force		Total heat, latent heat plus sensible heat above 0° Fahrenheit.
	In Inches.	In Atmospheres.			In Inches.	In Atmospheres.	
°Fah.							
32	0.1811	0.006	1123.70	248	58.7116	1.962	1189.58
50	0.3606	0.012	1129.10	266	79.9321	2.671	1194.98
68	0.6846	0.023	1134.68	284	106.9930	3.576	1200.56
86	1.2421	0.042	1140.16	302	140.9930	4.712	1205.96
104	2.1618	0.072	1145.66	320	183.1342	6.120	1211.54
122	3.6212	0.121	1151.06	338	234.7105	7.844	1216.94
140	5.8578	0.196	1156.64	356	297.1013	9.929	1222.52
158	9.1767	0.306	1162.04	374	371.7590	12.425	1227.92
176	13.9621	0.466	1167.62	392	460.1943	15.380	1233.50
194	20.6869	0.691	1173.02	410	560.9673	18.848	1238.90
212	29.9212	1.000	1178.60	428	684.6584	22.882	1244.48
230	42.3374	1.415	1184.00	446	823.8723	27.535	1249.88

MECHANICAL EFFECT.

We will now take into consideration the mechanical effect of steam, and a common-place demonstration will serve our purpose.

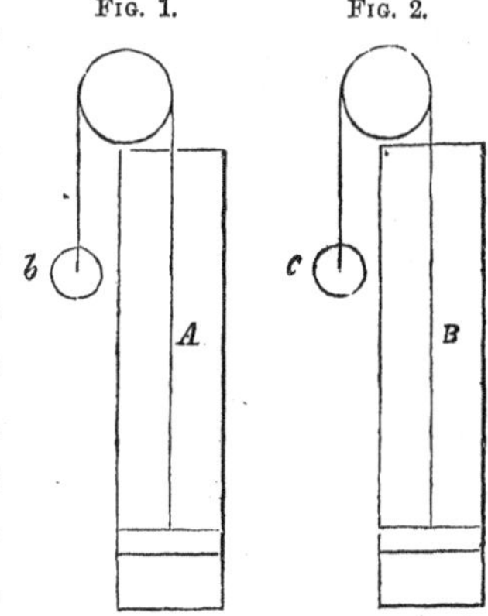

Suppose a cylinder, A, Fig. 1, to be one square inch in area of cross section, and fitted with a steam tight piston, attached by means of a flexible cord to the weight *b*, which is of sufficient size to balance the weight of the piston, and all the parts to work without friction. Now suppose a quantity of water, equal to one cubic inch, to be placed in the bottom of this cylinder, and a fire to be lighted under it. The temperature of the water will gradually rise until it attains 212°, when it will commence to boil, and the piston will soon begin, and continue to rise—if the cylinder be long enough—until it obtains a height of 1700 inches from the base. This 1700 is the volume of steam at atmospheric pressure, the water being 1, from which it is generated. If, now, we suppose to be added to the weight, *b*, another weight equal to the pressure of the atmosphere—or a fraction less, so that motion may ensue—and the steam under the piston to be condensed, the piston will return to the bottom of the cylinder by the pressure of the atmosphere, through a space of 1700 inches, and will have raised the extra weight of 14.7 lbs. appended to *b*, up that distance. Hence this cubic inch of water, by its evaporation, produced a mechanical effect of raising 14.7 pounds through a space of 1700 inches $= (14.7 \times 1700) = 24{,}990$ pounds through one inch.

Let us now take another cylinder, B, Fig. 2, similar in every respect to A, excepting that the piston has a weight laid upon it equal to the pressure of the atmosphere, viz., 14.7 pounds, and suppose a fire to be lighted under this cylinder. The water, as in the other case, will be heated up to the boiling point,—which, in this case, will be 250°, corresponding to the pressure of two atmospheres—when it will commence to evaporate, and the piston will rise until it obtains a height of 900 inches from the base, this being the volume of steam under the pressure of two atmospheres, water being 1. If, now, we suppose this piston to be fixed where it is, the weight removed from the top of it and applied to *c*, then the steam condensed and the piston unfixed, it will return to the bottom of the cylinder, raising the weight applied to *c*, up a distance of 900 inches. Now, then, since the weight of 14.7 lbs. was first raised 900 inches on the top of the piston, and afterwards raised the same distance by being attached to *c*, the total distance moved $= (900 \times 2) = 1800$ inches, which is equal to $(14.7 \times 1800) = 26460$ pounds raised one inch. The difference, therefore, between the work done in the first and second case $= (26460 - 24990) = 1470$ lbs. raised one inch high, which is 5.88 per cent. of the first number. If this extra work was obtained without any extra fuel, which would be the case were the total heat in steam at all temperatures a constant quantity, it would be all gain, but as such is not the case, and as more heat is required in the latter than in the former case, we will see what this amounts to, and the difference between this loss and the other gain will show the true gain. In the first instance, it will be seen that the total heat in the steam was 1178.6°, and in the second, 1190°; hence, supposing the water in both cases to be at a temperature of 100° before the fires are lighted—which is about the temperature at which water is fed into marine boilers—there would be required in the latter case $(1190° - 100) = 1090°$ from the fuel, and in the former case $(1178.6° - 100°) = 1078.6°$ from the fuel; difference 11.4°, which is 1.057 per cent. of 1078.6°.

The extra fuel, therefore, required under the pressure of two atmospheres is 1.057 per cent., and the extra work done is 5.88 per cent., leaving a gain of 4.823 per cent.

In the same way we could ascertain what the gain would be at any other pressure, either higher or lower; but these examples suffice to show that the higher the pressure of the steam, the greater is the mechanical effect with the same amount of fuel, but the gain is small, and in practice, therefore, where great accuracy is not required, it is neglected altogether.

Starting, therefore, from the assumption that the mechanical effect performed by the same amount of fuel is the same, no matter what the pressure may be under which the steam is generated, we shall proceed to the study of the

EXPANSION OF STEAM.

Opening a communication with the cylinder and shutting it off again before the piston arrives at the end of the stroke is called expansion of steam, or working steam expansively. Thus, supposing steam to be admitted into the cylinder until the piston arrives at half stroke, and the communication then to be shut off, the steam already in the cylinder, by its expansion, will force the piston to the end of the stroke; by which arrangement we gain all the work performed after the steam is cut-off.

Fig. 3.

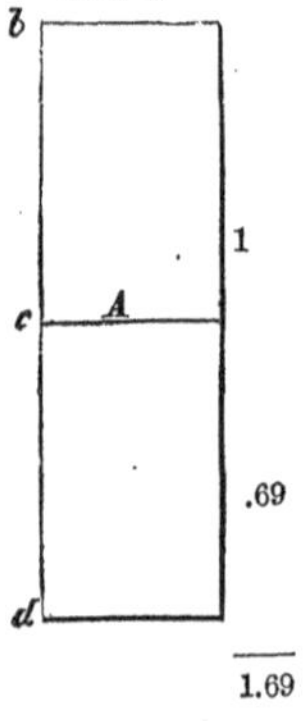

b, c, A, d, 1, .69, 1.69

Take, for instance, a cylinder, A, Fig. 3, two units in length, one unit in area of cross section, and an initial pressure of 1, the work performed during the first half stroke, *i. e.*, while the piston travels from *b* to *c*, will be $1 \times 1 \times 1$ (area $\times$ pressure $\times$ distance travelled $=$) 1, and the work performed during the latter half stroke $=$ $1 \times .69 \times 1 = .69$, the total work, therefore, performed throughout the stroke $= 1.69$. Now, if the steam, instead of being expanded from *c* to *d*, had been exhausted at *c*, the total work performed would have been only 1 instead of 1.69, and the quantity of steam would have been the same, hence we see that by cutting off at one half the same steam performs 69 per cent. more work. This 69 per cent. is what is termed the gain in cutting off, but it does not, however, represent the saving in fuel, as we will show presently; but before proceeding to illustrate that subject we will explain to the student, from what source we derive this 69.

Marriotte's law of gases is, that the spaces occupied are inversely as the pressures. That is to say, if steam of 20 pounds pressure per square inch, be allowed tc expand into double the space, the pressure will be 10 lbs.; if triple, $6\frac{2}{3}$ lbs.; if four times, 5 lbs.; if five times, 4 lbs., and so on. This theory would be literally correct did the temperatures remain constant; but as the temperature of all gases becomes reduced by expansion, the law does not hold good; nevertheless, in the steam engine, where there are so many extraneous circumstances which practically affect all calculations appertaining to the same, it is considered all that is ever required, and from its extreme simplicity is universally adopted.

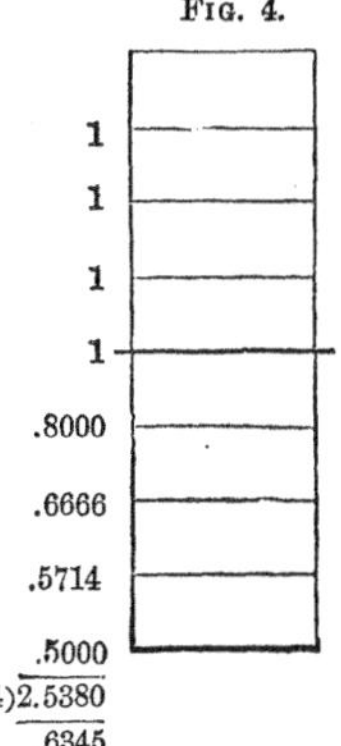

Fig. 4.

From this law the pressure can be ascertained approximately by dividing the cylinder into a number of equal parts, say eight, ascertaining the pressure at each of those points, and taking the mean. If the initial pressure, as before, be supposed to be unity, the pressure at each of the first four divisions cutting off at half stroke will be 1; at the fifth division ($\frac{4}{5}=$) .8; at the 6th ($\frac{4}{6}$) $=$ .6666; at the 7th ($\frac{4}{7}=$) .5714; at the 8th ($\frac{4}{8}=$) .5; the mean pressure, therefore, by this process, after the steam is cut off $=.6345$, and the mean pressure before it is cut off $=1$, the mean, therefore, throughout the stroke $=\left(\frac{1+.6345}{2}\right)=.8172$.— This, however, is only an approximation, and in order to arrive at any degree of accuracy, the divisions would have to be very numerous, which would render the operation tedious and lengthy. Fortunately, however, we can dispense with this part of the calculation altogether, for the Naperian or Hyperbolic logarithms, as set forth in the following table, furnish to our hand the ratios of pressures:

TABLE OF HYPERBOLIC LOGARITHMS.

Number.	Hyperbolic Log.	Number.	Hyperbolic Log.	Number.	Hyperbolic Log.	Number.	Hyperbolic Log.	Number.	Hyperbolic Log.
1.05	.049	3.05	1.115	5.05	1.619	7.05	1.953	9.05	2.203
1.1	.095	3.1	1.131	5.1	1.629	7.1	1.960	9.1	2.208
1.15	.140	3.15	1.147	5.15	1.639	7.15	1.967	9.15	2.214
1.2	.182	3.2	1.163	5.2	1.649	7.2	1.974	9.2	2.219
1.25	.223	3.25	1.179	5.25	1.658	7.25	1.981	9.25	2.225
1.3	.262	3.3	1.194	5.3	1.668	7.3	1.988	9.3	2.230
1.35	.300	3.35	1.209	5.35	1.677	7.35	1.995	9.35	2.235
1.4	.336	3.4	1.224	5.4	1.686	7.4	2.001	9.4	2.241
1.45	.372	3.45	1.238	5.45	1.696	7.45	2.008	9.45	2.246
1.5	.405	3.5	1.253	5.5	1.705	7.5	2.015	9.5	2.251
1.55	.438	3.55	1.267	5.55	1.714	7.55	2.022	9.55	2.257
1.6	.470	3.6	1.281	5.6	1.723	7.6	2.028	9.6	2.262
1.65	.500	3.65	1.295	5.65	1.732	7.65	2.035	9.65	2.267
1.7	.531	3.7	1.308	5.7	1.740	7.7	2.041	9.7	2.272
1.75	.560	3.75	1.322	5.75	1.749	7.75	2.048	9.75	2.277
1.8	.588	3.8	1.335	5.8	1.758	7.8	2.054	9.8	2.282
1.85	.615	3.85	1.348	5.85	1.766	7.85	2.061	9.85	2.287
1.9	.642	3.9	1.361	5.9	1.775	7.9	2.067	9.9	2.293
1.95	.668	3.95	1.374	5.95	1.783	7.95	2.073	9.95	2.298
2.	.693	4.	1.386	6.	1·792	8.	2.079	10.	2.303
2.05	.718	4.05	1.399	6.05	1.800	8.05	2.086	15.	2.708
2.1	.742	4.1	1.411	6.1	1.808	8.1	2.092	20.	2.996
2.15	.765	4.15	1.423	6.15	1.816	8.15	2.098	25.	3.219
2.2	.788	4.2	1.435	6.2	1.824	8.2	2.104	30.	3.401
2.25	.811	4.25	1.447	6.25	1.833	8.25	2.110	35.	3.555
2.3	.833	4.3	1.459	6.3	1.841	8.3	2.116	40.	3.689
2.35	.854	4.35	1.470	6.35	1.848	8.35	2.122	45.	3.807
2.4	.875	4.4	1.482	6.4	1.856	8.4	2.128	50.	3.912
2.45	.896	4.45	1.493	6.45	1.864	8.45	2.134	55.	4.007
2.5	.916	4.5	1.504	6.5	1.872	8.5	2.140	60.	4.094
2.55	.936	4.55	1.515	6.55	1.879	8.55	2.146	65.	4.174
2.6	.956	4.6	1.526	6.6	1.887	8.6	2.152	70.	4.248
2.65	.975	4.65	1.537	6.65	1.895	8.65	2.158	75.	4.317
2.7	.993	4.7	1.548	6.7	1.902	8.7	2.163	80.	4.382
2.75	1.012	4.75	1.558	6.75	1.910	8.75	2.169	85.	4.443
2.8	1.032	4.8	1.569	6.8	1.917	8.8	2.175	90.	4.500
2.85	1.047	4.85	1.579	6.85	1.924	8.85	2.180	95.	4.554
2.9	1.065	4.9	1.589	6.9	1.931	8.9	2.186	100.	4.605
2.95	1.082	4.95	1.599	6.95	1.939	8.95	2.192	1000.	6.908
3.	1.099	5.	1.609	7.	1.946	9.	2.197	10000.	9.210

The hyperbolic logarithm of any number can be found by multiplying the common logarithm by 2,30258509.

From the nature of hyperbolic logarithms they are thus very useful in working steam expansively.

Let the Line A, B, Fig. 5, represent the pressure of steam—which we will assume to be unity—at the time the cut-off valve closes; C, D, half the length of A, B, and the line A, C, a hyperbolic curve, .69+ from the table gives the mean length of all the ordinates, 1, 2, 3, 4, &c., which before we had to arrive at by approximation. If the cut-off valve, instead of closing at half stroke, had closed at some other point, say, when the piston had traveled only one-fourth its distance, C, D, would be one-fourth of *a*, *b*, and the curve A, C, would have extended from *a* to *c*, giving 1.38+ as a mean of all the ordinates below *a*, *b*.

FIG. 5.

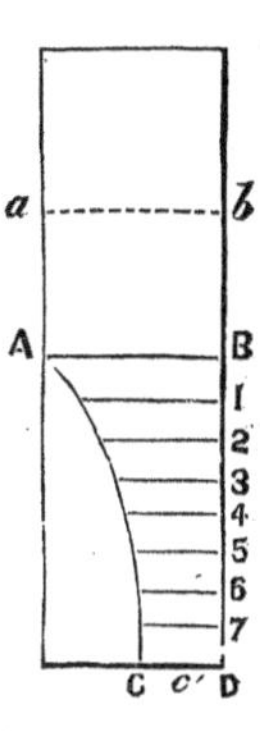

All we require then in working examples in expansion of steam, according to Marriotte's law, is to know the initial pressure and point of cutting off, from which we can deduce the mean pressure, pressure at the end of the stroke, percentage of gain, &c., by having before us a table of hyperbolic logarithms; but if it be required to make such calculations, when a table of this kind is not come-at-able, it can be done in the manner we have previously shown.

EXAMPLE 1st. Suppose you have a cylinder in which you are using steam of 20 pounds pressure per square inch, inclusive of the atmosphere, and cut off at half stroke, what is the mean pressure, pressure at the end of the stroke, and per centage of gain.

2

Ans. 1*st.* From the foregoing considerations we know that had the pressure of steam been 1 pound instead of 20, all we would have to do would be to take .69314 out of the table, add 1 to it and divide by 2; therefore to find the mean pressure we have this rule: *As the number of times the steam is expanded, is to the hyperbolic logarithm of that number plus* 1, *so is the initial to the mean pressure*, hence 2 : 1.69314 : : 20 : 16.9314lbs. mean pressure.

Ans. 2*d.* 20÷2=10 lbs. pressure at the end of the stroke.

Ans. 3*d.* Work performed before expansion, 1. after expansion, .69314. Therefore 1 : .69314 : : 100 : 69.314 per cent. gain by cutting off at half stroke.

Back Pressure.

Inasmuch as it is impossible in practice to obtain a perfect vacuum, there is always a certain amount of steam in the cylinder opposed to the motion of the piston, and this is termed *back pressure.* Suppose for example, there was in the above instance 4 lbs. per square inch back pressure, the mean *effective*, or unbalanced pressure, would be 16.9314−4=12.9314 lbs., and the unbalanced pressure at the end of the stroke would be 10−4=6 lbs.

Example 2d. Suppose the steam in example 1st had been cut off at a ¼ from commencement of stroke, what would have been the mean pressure, pressure at the end, and percentage of gain in that case? Also the mean unbalanced pressure, and unbalanced pressure at the end, the back pressure being 4 lbs per square inch?

Ans. 1*st.* 4 : 2.38629 : : 20 : 11.93145 lbs. mean pressure.

Ans. 2*d.* $20 \div 4 = 5$ lbs. pressure at the end.

Ans. 3*d.* 1 : 1.38629 : : 100 : 138.629 per cent.

Ans. 4*th.* $11.93145 - 4 = 7.93145$ mean unbalanced pressure.

Ans. 5*th.* $5 - 4 = 1$ lb. unbalanced pressure at the end.

It is useless here to multiply examples; those already given we consider sufficient to give the student a clear understanding of the manner in which these calculations are made, but we would recommend him to make a number of others for himself, and work them out so as to render himself the more familiar and ready with the *modus operandi.*

We come now to the percentage of gain of fuel by using steam expansively. It has been previously shown that when the steam is cut off at one half, the work done before expansion takes place being represented by 1, the work done afterwards is .69; the total work therefore performed is 1.69; now had not the cut-off valve closed at all, the total work performed would have been 2. Hence by this operation we have the power of the engine reduced from 2 to 1.69. It is therefore necessary to increase the initial pressure of steam to make up this decreased power; and keeping in view Marriotte's law, we will let this pressure be represented by x; hence 2 : 1.69 : : x : .845 x, the mean pressure. It is manifest that the mean pressure in this case must be the same as if the steam followed full stroke, in order that the powers may be the same; consequently

$$.845\,x = 1 \text{ lb.}$$

$$x = 1.18 \text{ lb. initial pressure.}$$

But only half a cylinder full of this steam is used to every full cylinder of the other, consequently the difference between $1.18 \div 2$ and 1, equals the saving, which is 41 per cent.

To ascertain the saving of steam at any other point of cutting off, take the hyperbolic log. of 3, 4, 5, 6, &c. as the cutting-off point may be $\frac{1}{3}$, $\frac{1}{4}$, $\frac{1}{5}$, $\frac{1}{6}$, &c., and proceed in the same manner, or, in other words, divide the whole length of the stroke by the portion traveled before the steam is cut off; take the hyperbolic log. of the quotient and proceed as above.

The 41 per cent. in the above example, is the saving in steam—that is to say, should a steamer, using steam full stroke, perform a certain distance in a certain time, cut the steam off at half-stroke, and increase the initial pressure in the ratio of 1 to 1.18, she would perform the same distance in the same time with 41 per cent. less steam. Not 41 per cent. less coals put into the furnaces, but 41 per cent. of that which reaches the cylinders minus the loss from condensation due to expansion, *i. e.* that portion of the fuel not combustible, and that portion passing out of the chimney in the shape of heat to produce draft, together with the loss from radiation and condensation before the steam reachers the cylinders must first be deducted. When this is done it will be found that the actual saving will be reduced to less than 20 per cent., which is about the real saving of fuel in practice cutting off at half stroke, and pro rata for any other point—varying somewhat according to better or worse constructions. Any engineer can satisfy himself on this point by using his steam with and without expansion for a sufficient given time, carefully weighing all the coals and recording all the data.

This should not therefore be confounded with the 69 per cent., which is theoretically the increased work performed by the same steam over what it would have performed had it not been cut off at all.

Expansion Valves.

There are a variety of expansion valves and arrangements for cutting off steam; the principles operating the more important of which we will now proceed to examine. The following diagrams or sketches will serve our purpose. We shall simply explain the leading features of each, in order to give an understanding of the principles that govern them, leaving the student to suggest for himself the alterations in the mechanical arrangement to adapt them to different types and arrangements of engines.

Fig. 1.

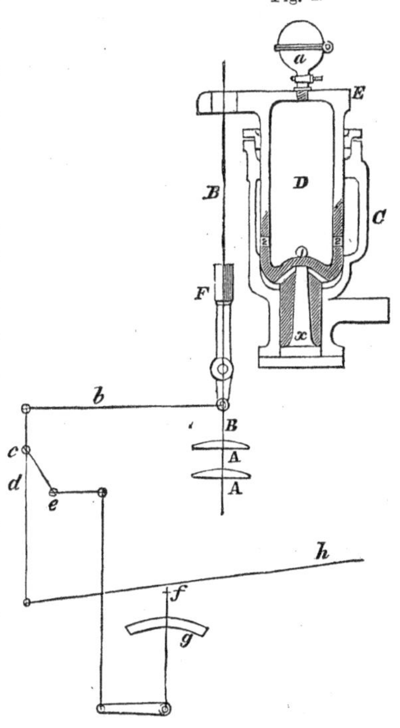

Figure 1 is a diagram of Sickel's momentarily adjustable cut-off, in which A, A, is the steam valve of the double poppet construction; B, B, valve stem; C, dash-pot, filled with water up to the line 1, 2; D, plunger, fitting in the dash-pot; E, stuffing-box, which is packed air and water-tight; *x*, hole in the bottom of the plunger D, to allow the water to enter when the plunger strikes it; *b*, a rod communicating motion to the wiper F, which trips the valve; *h*, a rod receiving motion from the air-pump beam or any other part having motion coincident with that of the piston. The motion of *h* is communicated through the vertical rod, having *c* as a fixed centre to *b*, and thence to the wiper F. The manner in which this cut-off operates is this: The valve stem, instead of being permanently attached to the lifting rod, is secured to it by a clutch and spring. The valve is lifted by the eccentric, (operating as in other cases,) but before it reaches its seat again, the wiper F, which vibrates back and forth, strikes the clutch, and detaches the valve from the lifter; the valve then, from the action of gravity, would fall, and strike its seat with a heavy blow; to prevent which, and allow the valve at the same time to fall quickly, it is attached to the plunger D, working in the dash-pot C. By this arrangement, before the valve reaches its seat, the plunger D strikes upon the water in the lower part of the dash-pot C, which is called the secondary reservoir, and thereby allows the valve to close without slamming, the water escaping into the cavity *x*, and also around the plunger, into the upper or primary reservoir. The plunger D, being hollow, small holes are bored into it in the vicinity of the line 1, 2, to allow the water to escape into it also.

We see that the cutting-off is effected by the wiper F tripping the valve; the sooner therefore the valve is tripped, the sooner the steam will be cut off. Now the manner in which this is made an adjustable cut-off, is accomplished by moving the handle *f* backward or forward on the arc *g*, which will move the centre *c* to one side or the other of its present position, the center *e* remaining constantly fixed, and therefore giving the wiper F a greater or less distance to travel before striking the clutch. By this means the cutting-off point can be varied for any part of the stroke. The handle *f* can be put in such a position that the valve will not be tripped at all, or it can be placed so that the valve will not lift at all, being exactly in the vertical position when the lifter commences to rise. The engine can be thus stopped by this cut-off. For the other end of the cylinder there is another dash-pot, &c., similar to the one described, the wiper being operated by a rod similar to *b*, attached to the center *d*.

Should there be too much water in the dash-pot, the valve will not seat quickly, but "hang," as it is technically termed. At *a* there is a cock for the purpose of supplying it with water. Attached to the dash-pot, there is usually another cock or valve for the purpose of letting out any superfluous water. Insufficient water is evidenced by the slamming of the valve.

This cut-off was formerly made without the wiper F, there being used instead a sliding cam, shaped something like this ◁. As the valve rose up, the clutch struck the bevel on this cam, which forced the clutch out of its position, and allowed the valve to fall. With this arrangement, however, it will be seen that the valve must trip while it is *rising*, and as it is at its highest position when the piston is about half stroke, it cannot be possible to cut off by this mode longer than half stroke; but with the arrangement of the wiper, it will be seen, inasmuch as it vibrates back and forth, that the valve can be just as well tripped on its descent as when rising, and this is the reason why it was substituted for the cam.

"*Stevens*."—The next cut-off that we shall take

into consideration is Stevens's, a diagram of which is shown in figure 2. A A are the steam toes; B B, the steam-lifting toes; D, rock-shaft arms; C C, the valves; *a*, pin in rock-shaft arm for eccentric hook. The manner in which this is made an adjustable cut-off, is by raising or lowering the toes A A, thereby giving them more or less lost motion. In the position in which they are shown in the diagram, it will be seen that they will have to travel a considerable distance before touching the toes B, B, and as the piston is in motion during this time; and the steam valve closed, the steam will be acting expansively. If the end of the toes A A be dropped lower down, the steam will be cut off shorter; if raised higher up, longer. By dropping the toes down, however, we diminish the lift of the valve, and also alter the lead. To retain the one, we raise the pin *a* in the rock-shaft arm, and the other we turn the eccentric a little ahead. To alter the point of cut-off, therefore, while the engine is in motion, so as to cut off shorter, we have first to drop the toes A A, then raise the pin *a*, and set the eccentric ahead. To cut off longer, reverse the operation.

Fig. 2.

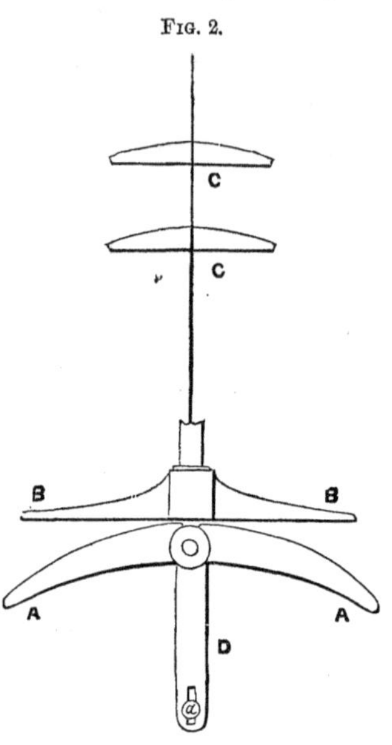

The number of things required to be altered in changing the point of cutting-off is a very great objection to this arrangement. In practice it has seldom been accomplished without stopping the engine.

Allen and Wells.—This cut-off is represented by sketch, figure 3. A, A, are the exhaust toes; B B, steam toes; C C lifting toes; D D, the valves; E E',

FIG. 3.

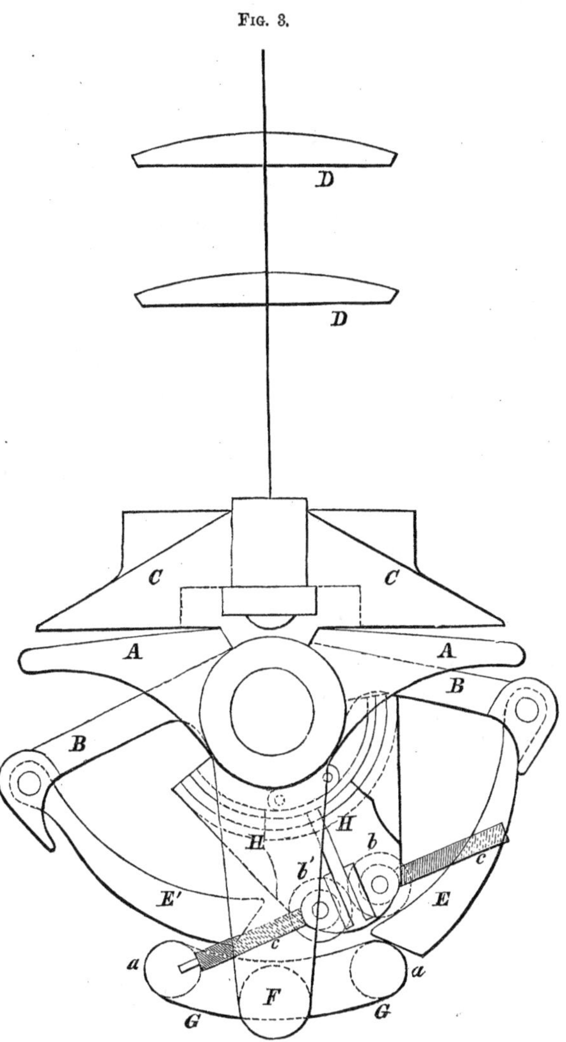

palls fitted to the end of the toes B, B'; F, rock-shaft arm, which is operated from the eccentric in the usual way; G G, a cross arm secured to the end of the rock-shaft arm; *a a*, rollers on the end of the cross-arm G, G′; H H, two arms fitted loosely on the rock-shaft. These arms receive their motion from any part of the engine having motion nearly coincident with that of the piston; *b b′*, rollers on these arms. This cut-off operates thus: The rock-shaft is put in motion by the eccentric. The pall E resting upon the roller *a*, is raised, and with it the toe B, and lifter toe C; but after the pall E is raised up so as to clear the roller *b*, the pall E slides in on top of *b*, which, having a downward motion, lowers the valve, while the rock-shaft arm continues to rise. The rollers *b b′*, being attached to the arms H H, which having motion nearly coincident with that of the piston, start to go down at nearly the same time the rock-shaft arm starts to rise. Now then by turning around the right and left-hand screw *c c*, the rollers *b b′*, will be set further apart, or closer together, and will therefore alter the time they will clear the end of the pall E, and hence the point of cutting off. To follow farther separate the rollers *b b′*, to cut off shorter, screw them closer together. In altering the point of cutting off we have nothing to do but to turn around the screw *c c*.

This cut-off is like "Sickel's," momentarily adjustable, but it cannot, however, be made to cut off quite so short as "Sickel's."

SLIDE CUT-OFFS.

In the use of the ordinary three-ported slide-valve, or other slide-valves combining both the steam and exhaust, the expansive principle can be carried only to a very small extent, owing to the derangement of the exhaust passages. Suppose, for instance, that sufficient lap be given to the steam side of the valve to cause the steam to be shut off at half-stroke, and suppose the same amount of lap be given also to the exhaust side, it is manifest, that when the steam is shut off, the exhaust will be shut off also, and the pent up steam, therefore, having no escape, and increasing in pressure as the piston approaches the end of the stroke, will act as a serious retarding force. This arrangement, therefore, cannot operate.

Now, then, suppose that we put lap on the steam side, as before, but none on the exhaust, in which event another difficulty equally great presents itself. It is this:

Supposing the valve, Fig. 4, to have neither lap nor lead, when the end *a* arrives at *a′*, steam will just begin to be admitted into the cylinder, but the point

FIG. 4.

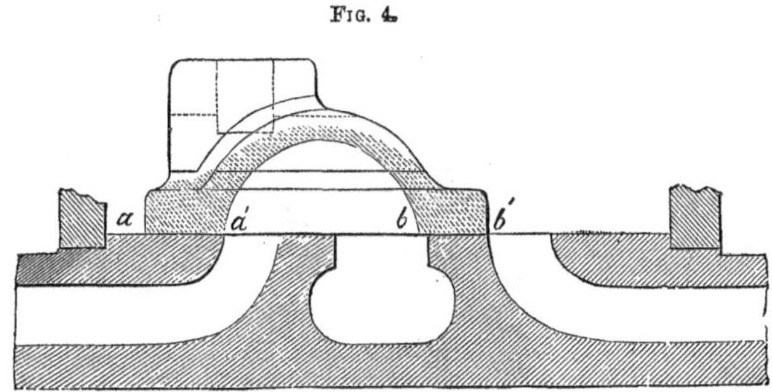

b, at the same time, will have arrived at the point *b′*, and steam just begin also to exhaust; now, then, let half an inch be added to each end of the valve at *a* and *b*, when the valve begins to open to steam in this case, *a*, instead of being at *a′*, will be half an inch past it; and, as there has been no lap added to the exhaust side, *b* will be half an inch past *b′*, so that the exhaust must have opened considerably before the piston arrived at the end of the stroke; hence, in this case, we exhaust too soon.

All we can do, therefore, in practice, is to strike a mean between these evils; that is to say, when we add lap to the steam side, add lap also to the exhaust side, but not so much so that we open the exhaust before the piston arrives at the end, and close it again before it reaches the other end. The shortest this kind of valve can be made to cut-off to advantage in practice, is considered about $\frac{2}{3}$ from commencement of stroke; but even this we consider most too short for beneficial working of large engines.

Owing to these confined limits, the beneficial results obtainable from the expansive principle by this arrangement is very small, which has led to the adoption of an independent slide cut-off valve, situated on a separate face, back of the steam valve, as shown in

Fig. 5.

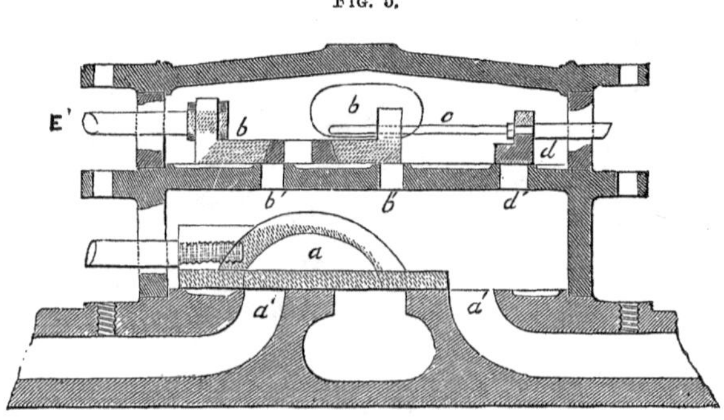

Fig. 5, in which *a'* is the steam, and *b b* the cut-off valve. The valve *a* having only sufficient lap to cover the ports *a' a'* fairly, when it is in the middle of the stroke, operates as in other cases, but the lap on *b b* can be made to any required extent, so that during a large part of the stroke the ports *b′ b′* are closed, preventing further access of steam to the cylinder, notwithstanding the steam valve itself is open. The valve *b b* is operated by an independent eccentric, through the valve stem E. In the position shown in the figure the steam is cut off about half stroke: *d′* shows another opening covered with the valve *d*, having a stem *c* sliding loosely through the valve *b b;* the other end of the stem passing through the chest, has a handle attached to it for the purpose of moving the valve *d*, in order to open the port *d′*, when the engine is stopped. This is necessary, for the reason that the engine may stop when the valve *b b* is in such a position as to prevent the steam from entering to the steam valve *a*, and the engine could not, therefore, be started. In the figure, the cut-off valve has but two ports for the admission of steam, but any number of ports can be made—the more numerous, the less stroke will be required to get the necessary opening. This is what is termed the gridiron valve, from the resemblance it bears to that very *useful* instrument.

After this valve is once made, the point of cutting off usually remains fixed, but it can, however, be varied

Fig. 6.

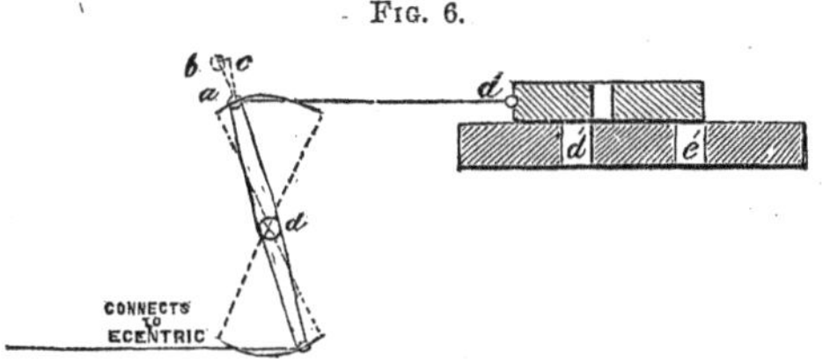

within narrow limits by altering the stroke of the valve. Thus, in Fig. 6, supposing the end of the valve stem to be raised from *a* to *b*, the valve, instead of being closed, as shown, will be open the distance *b c*, and will therefore have that much additional to travel before the steam is cut off; hence, by increasing the travel of the valve we increase the point of cutting off, and conversely, supposing the pin *a* had been lowered in the rock-shaft arm the distance *a e*, equal to *a b*, the ports, instead of being closed, as shown, would be closed the distance *b c;* the steam, therefore, would be cut off sooner. But by altering the point of cutting off we also alter the lead of the valve; for, taking the case in which we increased the travel of the valve, we see that when it would have been closed with the original lead, it lacked the distance *b c*. If its travel had been reduced, it would have lacked that much of being open. To obviate this, whenever the travel of the valve is altered, the eccentric should also be altered, so as to retain the original lead.

If the travel of the valve be made too great, the valve *d* will pass entirely over the port *d'*, and gradually close *e'*, unless they be set some distance apart. If the travel be made too small, the steam will be shut off, and the motion of the eccentric being reversed long before the piston arrives at the end of the stroke, steam will be admitted to it again before the steam valve closes.

From the above facts, and the figure before us, we draw the following general conclusions in reference to this kind of slide cut-off valves:

That, with a given amount of lap, the cutting off point can be varied from the longest point of cutting off allowable by said lap, to a certain point within the stroke, by reducing the stroke of the valve and altering the eccentric so as to retain the original lead. If the stroke be reduced beyond this, steam will be shut off and given to the piston again before it arrives at the end of the stroke. In practice, this variation will not amount to more than from about $\frac{5}{8}$ to $\frac{3}{8}$ of the stroke.

In altering the stroke of the valve, the slot through which the pin *a* moves should be an arc of a circle, struck with a radius equal to the length of the link *d a*, and with *d* as a centre.

With equal leads, the cutting off point cannot be effected equally on both ends of the cylinder with a slide valve, owing to the connecting rod acting out of parallelism, or, in other words, owing to the crank not being at 90° when the piston is half way. The shorter the connecting rod, the greater the discrepancy.

FIG. 6½.

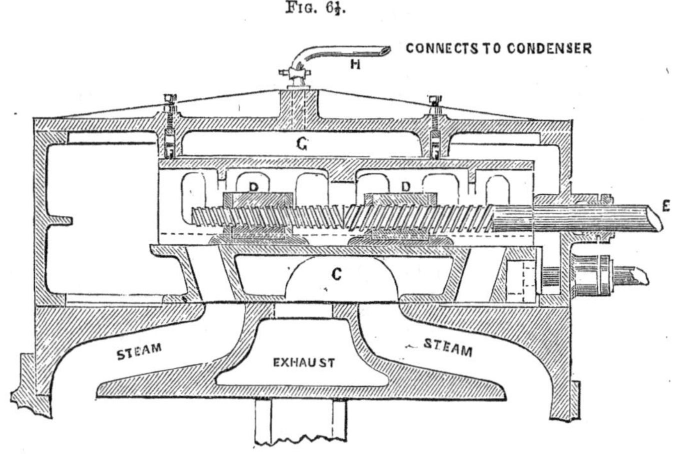

Fig. 6½, is an arrangement of cut-off valve as constructed by Messrs. Merrick & Son, of Philadelphia, in 1855, for the U. S. Steam Frigate "Wabash."

In consequence of the satisfactory manner in which it worked on board that vessel; its simplicity, and easy adjustment for cutting off at any portion of the stroke likely to be required, it has been applied to nearly all the U. S. Screw ships recently constructed, as also to a number of other engines. C is the steam valve; D D are the cut-off valves, attached to the valve stem E by right and left hand screws working in nuts let into the valves ; F F, rings in the steam chest cover, fitting close down on the back of the main steam valve, enclosing the space G, which is connected to the condenser by the pipe H, for the purpose of balancing the valve.

This cut-off can be worked by a separate eccentric, or from any part of the engine having a motion coincident with that of the piston.

To alter the point of cutting-off, a wheel is on the end of the valve stem E, which, if turned in one direction, will draw the valves closer together, and the openings will not be closed so soon, consequently the steam will follow the piston farther, *i. e.* cut off longer.

To cut off shorter, the operation is reversed.

OTHER KIND OF VALVES.

Having explained the principles of the leading cut-offs, we will now take a glance at some of the most prominent steam and exhaust valves now in use; but, inasmuch as the student is supposed to understand the leading features of most of these, we will not devote much time to this part of our study.

Figure 7 is a diagram of a double poppet valve, in which the rectangular space, *a b c d* is the opening to the cylinder; A B, the steam valves, and C D, the exhaust valves. The object of this arrangement is to make the valve a balance valve. Thus the steam acting on the top of A and bottom of B, if they were of equal size, an equilibrium would be established, but the valve B is made just small enough to slip through

Fig. 7.

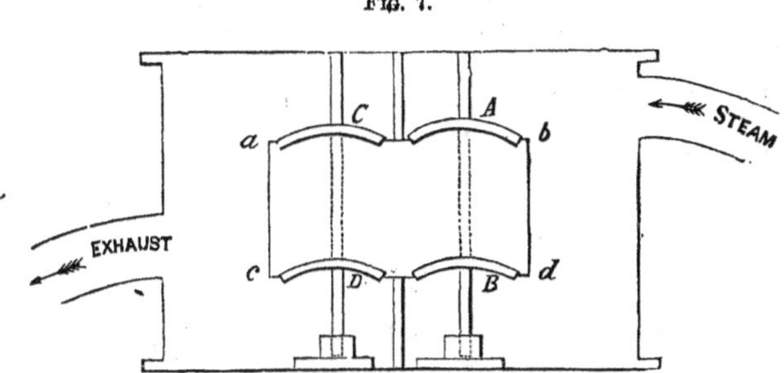

the upper seat, so that the difference in area serves to keep the valves fairly in their seats.* On the exhaust side the reverse is the case. The steam acts under C, and on top of D, the lower valve D is usually made the larger. In order to get D into its place, the upper seat is either made removable by being secured in its place by tap-bolts, or a hand-hole is cut in the side of the steam-chest, or, in some cases, it is passed in through the cylinder nozzle.

Figure 8 is a diagram of the single poppet valve, in which A is the steam valve, and B, the exhaust. With these kinds of valves we see that we require considerable power to operate them by hand, as we have the full pressure of steam on the back of A, and also the exhaust on B; but

Fig. 8.

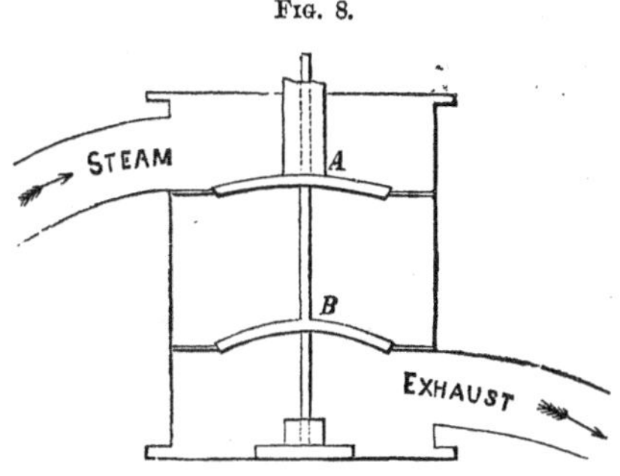

when the engine is hooked on the pressure is in part balanced. On the steam valve this is occasioned at the time the valve is opened, by the exhaust valve

* In some cases, the areas of the valves are equal, and they are seated by their own weight.

3

being closed before the piston arrives at the end of the stroke, producing the pressure called cushion. And on the exhaust valve the pressure is reduced (at the time the valve is opened) by expansion. In some cases this pressure is but little above that in the condenser. It is therefore obvious that these valves can be made to work with but little power from the engine. They also have the advantage of being easily made tight and occupying but little room.

The disadvantage of working by hand, however, led to the adoption of the double poppet valve, the single poppet being the earlier invention. The double poppet valve is the one now almost universally used in American low-pressure river, or marine paddle-wheel engines.

Figure 9 is a representation of what is termed "Hornblower's" valve, in which *a a b b* are the valve

Fig. 9.

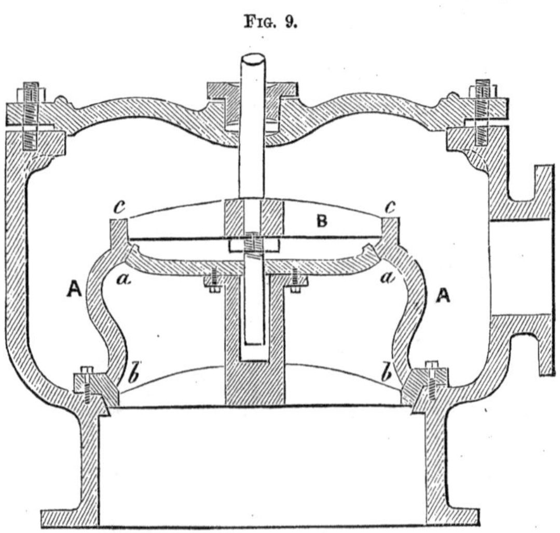

seats; A A, the valve; B, one of a number of crossbars secured to the top of the valve, to which the valve stem is attached. From the figure it will be seen that the only surface the steam has to act upon to keep the valve in its seat, is the upper edge, *c c*, of the valve; it is therefore an equilibrium valve.

Figure 10 is what is termed a box valve; *a a* are the parts communicating with the cylinder; *b*, steam-pipe; *c*, the exhaust; A A, the valve having an opening through its center communicating with the exhaust, *c*; *d d*, packing. An inspection of the figure will show the operation of the valve. The object of this kind of valve is also to establish an equilibrium.

Fig. 10.

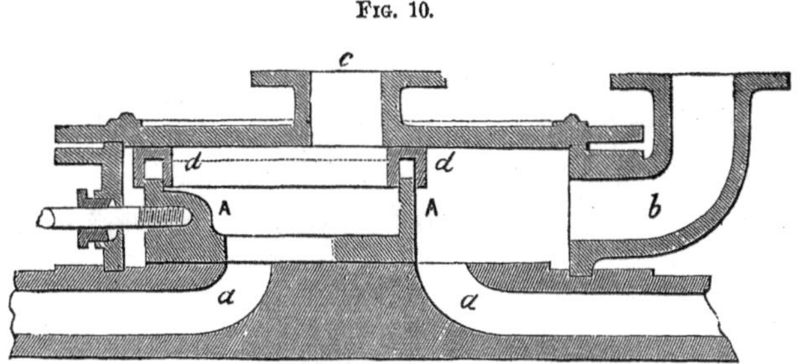

Figure 11 is a longitudinal section, and figure 12 a top view of what is termed the equilibrium slide.

Fig. 11

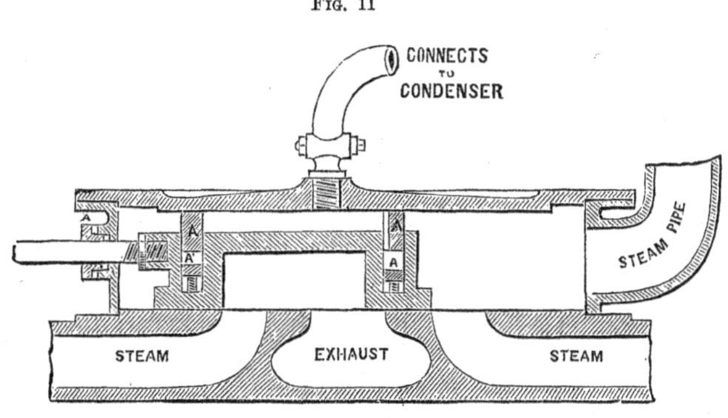

This valve has a ring, A A, on the back of it, which being made steam tight, the pressure is taken off the space enclosed by the ring. The pressure is taken off the back of nearly all the valves of large engines now-a-days, fitted with the short slide, either in this way or by having the ring secured to the top of the chest, and the valve sliding under it.

Fig. 12.

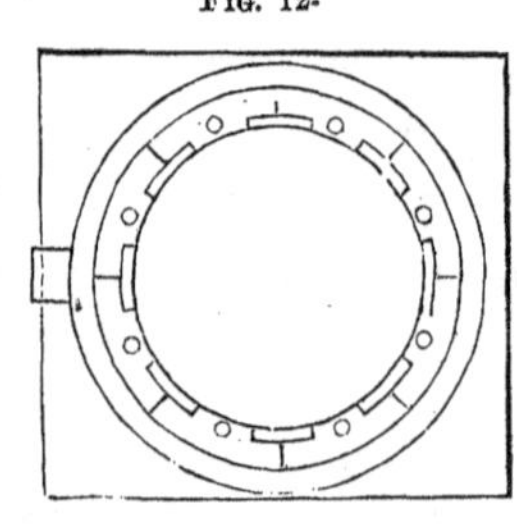

Figure 13 shows a slide valve A A A, having openings *b b* through it for the admission of steam;

Fig. 13.

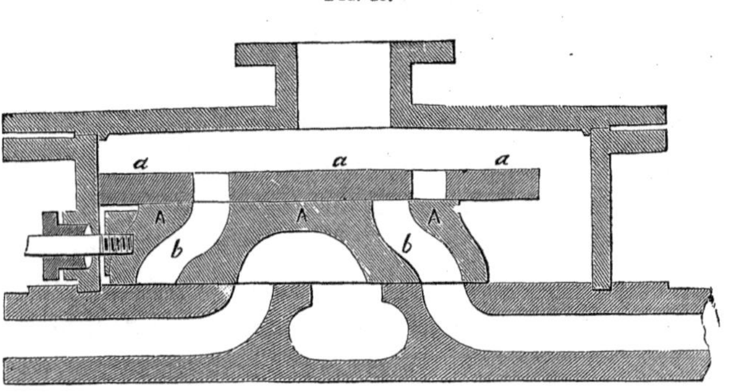

a a a is another valve sliding on the back of the valve A A A; *a a a* is the cut-off, which operates thus: The valve A A A being put in motion, and the cut-off valve lying loosely on its back, is carried with it until the end of the valve *a*, *a*, *a*, strikes the steam chest, when its motion is arrested, while the steam valve continues to move, the result is the closing the opening *b*, and the cutting off the steam. The sooner, therefore, the slide *a a a* strikes the chest, the sooner the steam is cut off. The point of cutting-off can be varied by having a screw running through the chest, which can be moved further in or out and against which the valve *a a a* strikes. With this arrangement it will be seen that the cut-off must close at furtherest a little before the piston arrives at half stroke, or not close at all. This cut-off is applicable to horizontal stationary engines.

Fig.14.

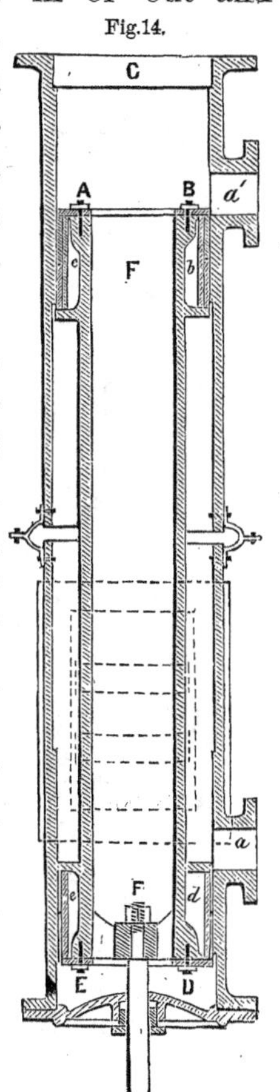

Fig. 14 is a piston valve, in which *a a′* are the openings into the cylinder; C, exhaust opening; A B D E the valve packed at *b c d e* with rings or other packing. In the position shown in the figure, steam is being exhausted through the openings *a′* and C into the condenser, while steam is being admitted into the opposite end of the cylinder through the opening *a*. When the valve has its full throw in the opposite direction, steam will be admitted through the opening *a′* while it is being exhausted through *a*, and the opening F F through the valve and through C into the condenser.

Figure 15 shows the long D slide, with the full

FIG. 15.

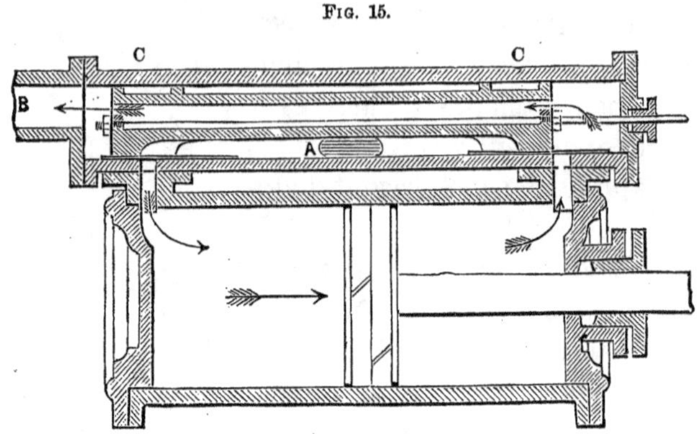

opening for steam under the piston; Fig. 16, same

FIG. 16.

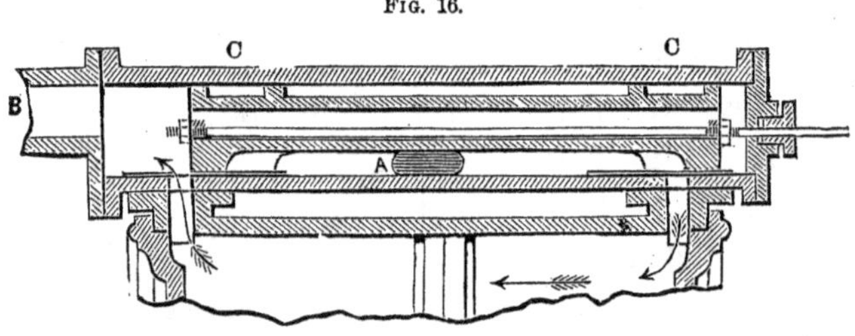

valve showing full opening for steam on top of the piston; Fig. 17, longitudinal section of the valve alone,

FIG. 18. FIG. 17.

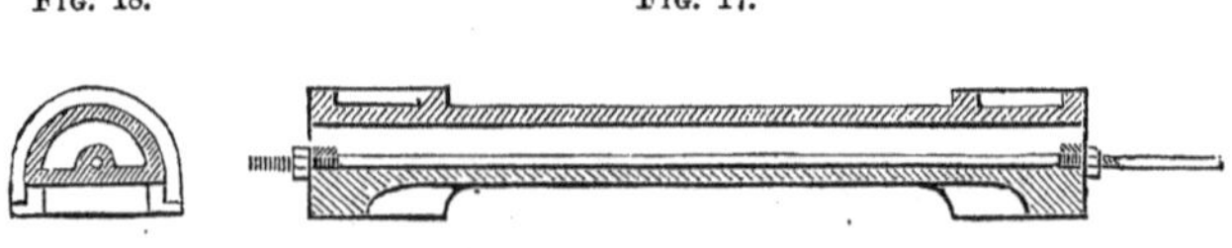

and Fig. 18, cross section of the same. A is the steam pipe, B, the exhaust, C, packing to keep the steam and exhaust separate, steam being admitted into the chest or valve casing at A, fills the vacant space under and around the valve, but cannot escape past the ends owing to the packing C C; and, when the valve is placed in the position shown in Fig. 15, steam is admitted under the piston in the direction shown by the arrows, at the same time that it is exhausted through the upper opening, and—the valve being hollow—through it and pipe B into the condenser. When the valve is moved in the opposite direction, steam is admitted above the piston in the direction shown by the arrows in Fig. 16, and exhausted through the lower opening directly through the pipe B to the condenser.

Fig. 19

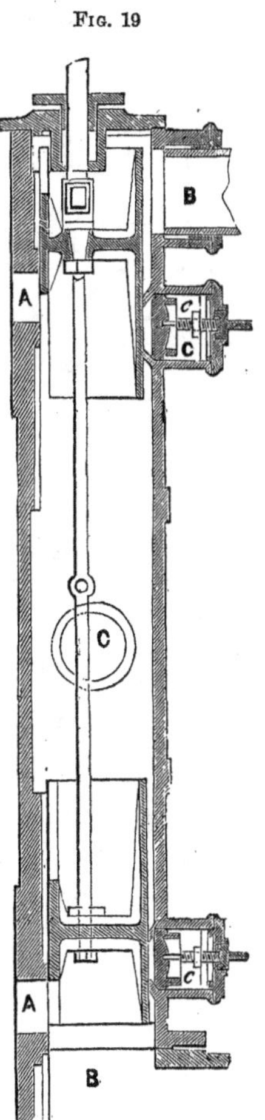

This style of valve is in extensive use on English marine, and other engines. The objection to it is the friction, requiring several men to work the starting bar when the engine is operated by hand.

Fig. 19 is a longitudinal section of the short D slide, and Fig. 20, an end view of the same. A A′ are the openings into the cylinder; B B, the communications to the condenser; C, steam pipe. In the position shown in the figure, steam is being admitted through A into the cylinder, and exhausted through A′ into the condenser, *c c* is packing on the back of the valve.

Fig. 20.

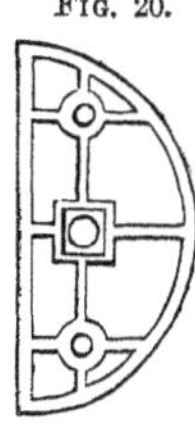

Figure 21 is a view of the Worthington pump steam valve; figure 22, the valve face, and figure 23, the valve seat. The figures explain themselves. In the ordinary slide valve, when it is moved in one direction, steam is given to the piston in the same direction, but the object of this valve, as invented by H. R. Worthington, of N. York, is to cause it, when moved in one direction, to give steam to the piston in the opposite direction. The valve being operated by an arm projecting from the piston rod, which strikes collars on the valve stem, renders it necessary that when the valve is moved in one direction, steam should be given to the piston in the opposite direction, in order to reverse its motion; by this arrangement the intervention of levers is unnecessary, as the end is accomplished direct.

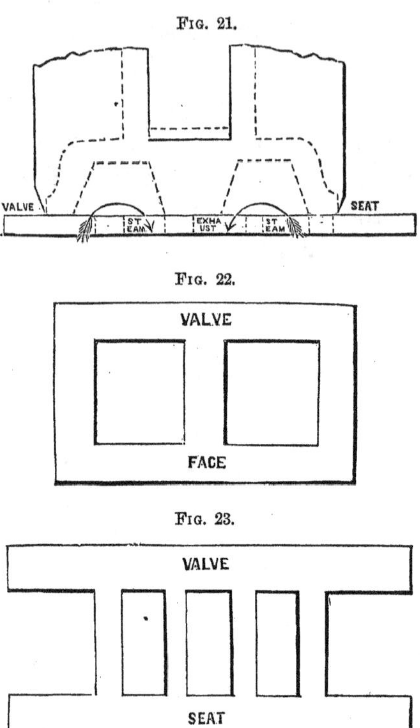

Fig. 21.

Fig. 22.

Fig. 23.

The Pittsburg Cam.—Figures 24, 25, and 26, show different forms of this cam. Like letters refer to like parts. A B C D is a yoke fitting over the cam *a b c*; E is a rod attached to the valve stem. F, main shaft of the engine to which the cam is secured. It will be seen that by the revolution of the cam *a b c*, within the yoke A B C D, the rod E will be caused to move back and forth, and thereby open and shut the valve.

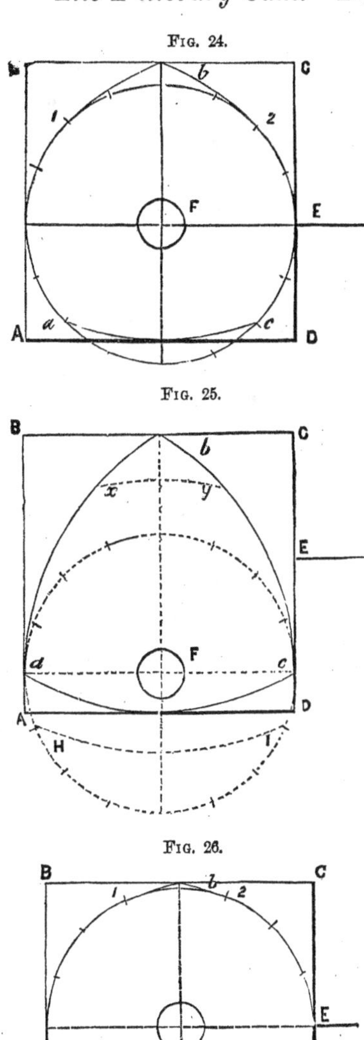

Fig. 24 is a cam made to cut off at half stroke; figure 26, ¼ stroke, and figure 25 follows full stroke. The manner in which these cams are laid off is this. From the centre F, with a radius dependent upon the stroke of the valve, describe a circle, as shown partly in dots and partly in full lines in the figures; divide this circle into any convenient even number of equal parts, say sixteen; then, supposing we wish to cut off at half stroke, taking figure 24, place one foot of the dividers having a radius equal to the diameter of the circle at C, and describe the arc terminating at *b*, then move the foot of the dividers from *c* to *a*, and describe another arc terminating also at *b;* then, with the same radius, and *b* as a centre, describe the arc *a c;* the figure thus enclosed will be the required cam. It will be observed that, while the cam is traveling the distance *a* 1—that being an arc of a true circle—no motion can be given to the valve, but while it travels from 1 to 2 the valve is opened and shut. Now, then, inasmuch as the piston moves from one end of the cylinder to the other for each semi-revolution of the cam, and inasmuch as the distance from *a* to 1 is the same as from 1 to 2, the valve remains necessarily closed during one-half of the stroke.

In figure 25, as no part of the outline of the cam is concentric to the shaft F, the valve must be in motion, all the time the cam is in motion. In figure 26, as three-quarters of the semi-periphery of the cam is concentric to the shaft F, the valve will remain closed during three-quarters of the stroke. Instead of making the points *b* sharp, as shown in the figures, they can be turned off, and, to retain the same dimensions on the cam, an equal amount added to the arc *a c*. Thus, taking figure 25, suppose we cut off the point of the cam to *x y*, and increase the lower extremity to H I, this will not alter the point of cutting off, but it reduces the travel of the valve, and has the effect of keeping the valve stationary when wide open, while the cam travels through the arc *x y*.

CHAPTER II.

THE INDICATOR AND INDICATOR DIAGRAMS.

THE steam engine indicator is an instrument used for the purpose of exhibiting the performance of the steam engine. By its application to the steam cylinder we can ascertain the following particulars: Whether the valves are properly constructed and set; steam and exhaust passages of the right size; whether the piston or valves leak; the amount of vacuum or back pressure, and pressure of steam upon the piston; the power of the engine; power required to overcome its friction, and also to work any machinery attached to the same, &c. In truth, it is the stethoscope of the physician, revealing the internal working of the engine.

The following description of the instrument and cut, Fig. 27, we take from Paul Stillman's Treatise on the Indicator. The cut shows the style manufactured at the Novelty Iron Works, New York city:

A is a brass case enclosing a cylinder, into which a piston is nicely fitted. To the piston-rod a spiral spring is attached to resist the steam and vacuum when acting against it. B is a pencil attached to the piston rod. C is an arm attached to the case, and supporting a cylinder D, which may be caused to rotate back and forth—a part of a revolution in one direction, by means of a line or cord *e*, attached to a suitable part of the engine—and in the other by means of a strong watch spring within the cylinder D. Outside this cylinder is to be wound a paper, upon which a diagram will be made, by the combined action of the piston and paper cylinder, representing, by its area, the power exerted on one side of the piston during the whole revolution of the engine. *f f* are springs to secure the paper to the cylinder; *g* is a scale divided into parts corresponding to the pounds of pressure on the square inch. These divisions, for convenience of measuring the diagrams with a common rule, are generally made in some regular parts of an inch, as 8ths, 10ths, 12ths, 20ths, 30ths; *h* is a cock by means of and through which it is connected with the engine cylinder.

Fig. 27.

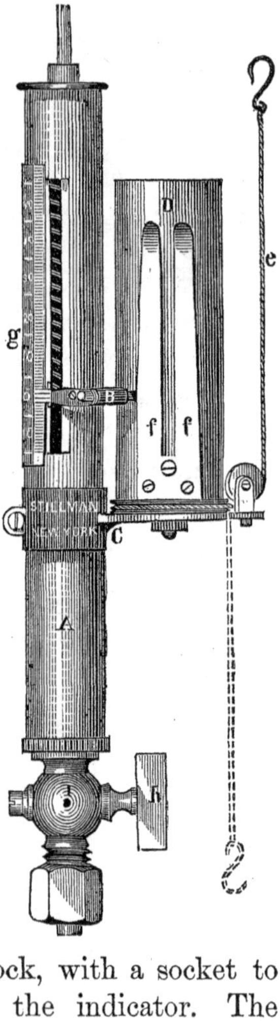

How to attach the Indicator.

Into whatever part of the engine it may be desired to apply the indicator, there must first be inserted a small stop-cock, with a socket to receive the one connected with the indicator. The instrument is to be set into this in such a position that the line attached to the paper cylinder shall lead through or over the guide pulley toward the place whence it is to receive its motion. An extension of this line should be connected with some part of the engine, the motion of which is coincident with that of the piston, and which would give the paper cylinder a motion of about three-fourths of a revolution. If the engine is of the construction denominated *beam* or *lever* engine, and is provided with a "parallel motion" the parallel bar, or a pulley on the radius shaft, furnishes the proper motion; if otherwise, the beam centre may be resorted to. In the kind denominated *square* engines, the centre of the air pump gives it. In horizontal and vertical *direct acting* engines, it will frequently be found necessary to erect a temporary rock-shaft, or lever, connected with the cross-head. Particular care should be taken, when the power of the engine is to be estimated, that the motion communicated be perfectly coincident with that of the piston.

In nearly all forms of the steam engine, the proper motion may be obtained by attaching a line to the cross-head, and passing it over a delicately constructed pulley, to the axis of which should be attached a smaller one, from which a line shall connect with the indicator. The proportional sizes of the two pulleys, of course, should be as the distance traveled by the piston to the length of motion given to the paper cylinder of the indicator. It will be necessary to attach a strong spring to the axis of these pulleys, to produce the reverse motion promptly. In an oscillating engine, it will be necessary that the indicator, with its fixtures, should be attached to the cylinder.

As the paper cylinder cannot make more than about three-fourths of a revolution without disturbing the point of the pencil, it will be seen that the line communicating the motion must be of a definite length. It also requires to be readily connected and disconnected.

The indicator having been attached to the steam cylinder, the paper secured smoothly on the cylinder D, figure 27, and the length of the line *e* being adjusted so that by the vibration of D it does not strike the stops, we will proceed to take a diagram, first taking care to see that the paper cylinder D is so fixed that the springs *f f* do not come in contact with the pencil B. The pencil B being adjusted so that it touches lightly on the paper, throw it back and attach the hook on the line E to the line receiving motion from the engine; then open the cock *h*, and allow the piston to work up and down several times, in order to heat and expand all the parts of the instrument. This being accomplished, turn the pencil on and take the diagram. Shut off the cock *h*, and apply the pencil again to the paper, and it will describe the atmospheric line.

Figure 28 is a diagram taken from the U. S. S.

Fig. 28

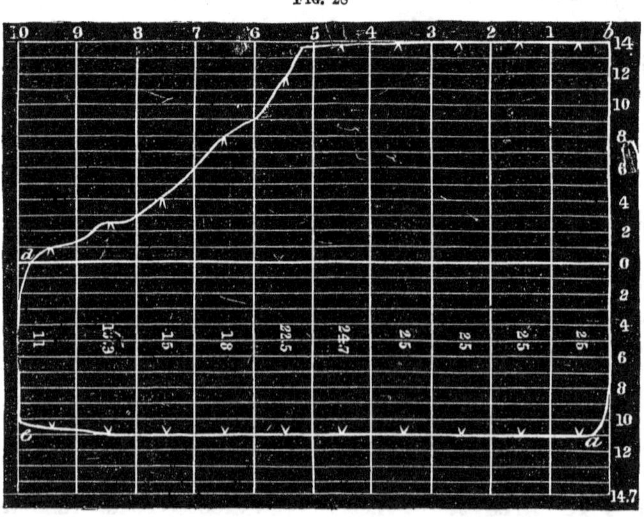

10) 204.5

20.45 lbs. mean unbalanced pressure.

Frigate "Powhatan," fitted with the double-poppet balanced valves, and Sickels' cut-off, on the 15th of January, 1854, while on the passage from Hong Kong, China, to the Loo-Choo Islands. At *a*, the piston of the indicator being at the bottom of its stroke, steam is admitted, forcing it up to *b*; at *b* the cylinder upon which the paper is wound—having motion coincident with that of the steam piston—starts to turn, describing the line *b c*; at *c* the expansion valve closes, and the pressure therefore gradually falls to *d*, where the exhaust valve opens and the pressure falls suddenly to *e*; the steam piston now starts on the return stroke, and the spring within the cylinder D, fig. 27, forces it back to the beginning *a* of the diagram. The line from *a* to *b* is called the receiving line; from *b* to *c* the steam line; from *c* to *d*, the expansion line; *d* to *e*, the exhaust; *e* to *a*, the vacuum line. The numbers in the vertical column on the right-hand side of the figure, are the pounds pressure; 14.7 is the true vacuum line, *o*, the atmospheric line, and 14, the initial pressure of steam above the atmosphere. The figures along the top line are the feet in length of the cylinder. It will be seen that the cut-off valve closed when the piston had traveled a very little beyond half stroke. The rounding at *d* and *a* is the lead and cushion on the exhaust. That is to say, the exhaust valve opened at *d*, before the piston arrived at the end of the stroke, and it also closed again at *a*, before the piston reached the other end of the stroke. Had there been no lead both of these corners would have been well defined.

In order to calculate the power of the engine, the mean pressure on the piston must be known, and from no source but the indicator can it be accurately ascertained. The manner of arriving at this is simply by taking the total pressure at different points and adding them up and taking the mean, as shown in fig. 28.

Figure 28 is what would be termed among engineers a good diagram; so is also figure 29, which we will take for a further elucidation of the subject.

Fig. 29.

Steam,...........	10
Vacuum,.........	27
Hot well,.........	106 Fahr.
Revolutions, ...	9.5
Throttle,.........	8.

"Powhatan" stb. cylinder, bottom
Nov. 7, 1855, 10 A. M.
One engine and one wheel in operation.
Smooth sea.

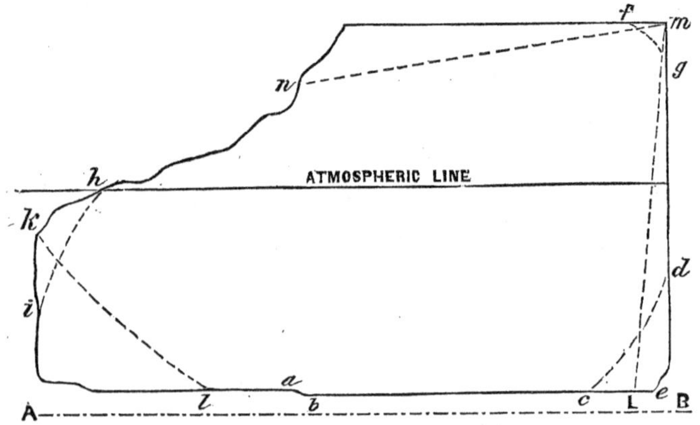

It appears from this diagram, however, that the piston of the indicator worked rather tightly, which occasioned it to stick a little in some places, as is evidenced by the steps in the expansion line, and also at *a b* in the vacuum line. If the piston of the indicator become much scratched, similar effects will be produced. Great care should, therefore, be taken in its use, to see that neither the piston works too tightly nor too loosely; for on the one hand it will stick, and thereby produce an imperfect outline, and on the other hand will produce the same effect by exhibiting false vacuum and expansion lines.

Should figure 29, instead of being as shown in full lines, have the lower right hand corner cut off as shown in dots at *c d*, the defect would have been that the exhaust valve closed too soon—at *c* instead of *e*—occasioning excessive cushioning. With some engines, however, a large amount of cushioning is necessary to prevent them from thumping on the centres.

Had the upper right-hand corner been rounding, as shown by the dotted line *f g*, the defect would have been that the steam valve opened too late. Had the exhaust corner been cut off, as shown by the dotted lines *h i*, the exhaust valve would have opened too soon; but had it been in the form shown by the dotted line *k l*, it would have opened too late, and after it did commence to open, would move with too slow a velocity, preventing the free escape of steam, or the exhaust passages would have been too small, which would produce a similar effect to the valve opening too slowly. Had the steam line, instead of being parallel to the atmospheric line, fallen down in the direction *m n*, it would have shown that the throttle was partially closed, or the steam passages too small, preventing the full flow of steam into the cylinder.

Should there be excessive lead given to the steam valve, the line *d m*, instead of being at right angles to the atmospheric line, will have the top inclined to the right as from L to M.

In taking a diagram for the purpose of estimating the power of the engine only, the atmospheric line is not necessary; but in order to ascertain the vacuum it cannot be dispensed with, unless the indicator piston be forced down to the perfect vacuum and held there until that line be described.

In a diagram taken from a non-condensing engine,

the atmospheric line will of course be entirely below it, owing to the back pressure occasioned from the passage of the exhaust steam through the openings and pipes. Had figure 29 been taken from a non-condensing engine, A B would have been the atmospheric line.

Figure 30 we have copied from Main and Brown's Treatise on the Indicator and Dynamometer. It was

Fig. 30.

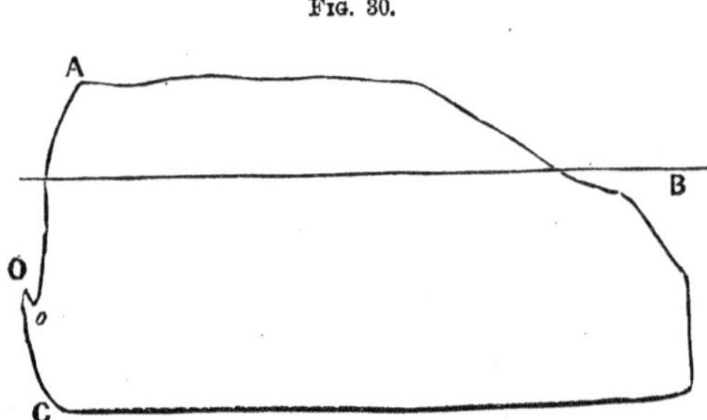

taken from an engine fitted with the long D slide. There are two defects exhibited in this diagram; the steam communication is opened too late and the exhaust too soon. At C the exhaust closes, causing the steam to be compressed to O, when the piston having arrived at the end of the stroke starts on its return, and the pressure falls to O′; at O′ steam is admitted, causing the line O′ A to be traced; at B the exhaust opens long before the piston arrives at the end of the stroke, allowing the steam to escape too soon. The hook, as shown at O, would only be made in very aggravated cases, where the steam is very much behind time.

Fig. 31 is obtained from the same source as figure 30. In this case the engine was working as a non-

FIG. 31.

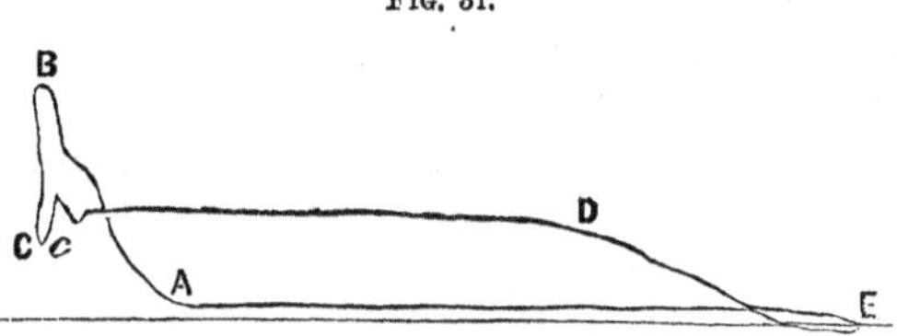

condensing engine with a very low pressure of steam. The exhaust closes at A, causing the pent-up steam to be compressed to B, where the steam valve opens, and the pressure in the cylinder being greater than that in the boiler, immediately falls to C. The hook at C is occasioned by the momentum of the indicator piston. At D the cut-off closes, causing the steam to be expanded to E, below the atmosphere. At E the exhaust valve opens and the pressure rises up equal to the back pressure, causing the loop on that corner of the diagram.

Figure 32 is a diagram drawn from memory, from one of a non-condensing engine that was once shown the author, with the request that he point out the defect in the engine from which such a diagram was taken. At first we did not see any reason why

FIG. 32.

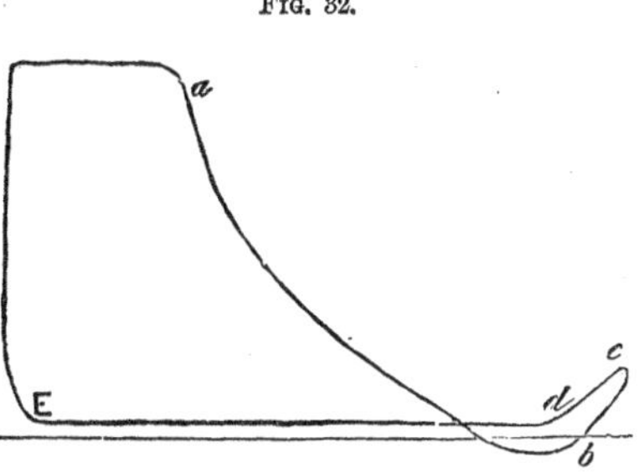

the pressure should rise from *b* to *c*, for supposing the exhaust to open at *b*, there could be no reason why the pressure should rise beyond *d*, the amount of back pressure on the opposite side of the piston. After looking at it a little closer, however, it occurred to us that such a diagram could be formed from a slide valve engine, and in this manner: Steam being admitted in the usual way until the piston arrived at *a*, the independent slide cut off the steam, whence it was expanded to the point *b*; at *b*—the steam valve having neither lap nor lead, and consequently still open—the cut-off again opened the communication with the cylinder, admitting fresh steam, which caused the line *b c* to be traced, partaking of the motion of the steam and piston. At *c* the steam is shut off by the steam valve itself, and the exhaust opened, the pressure therefore falls from that point to *d*, and the exhaust line is traced. In a non-condensing engine diagram, where of course there can be no vacuum line, the line from *c* to *e* inclusive is termed the exhaust.

A perfect Diagram.—According to the law laid down by Marriotte, which we have previously studied under the head of expansion of steam, the expansion curve of an indicator diagram should be a true hyper-

FIG. 33.

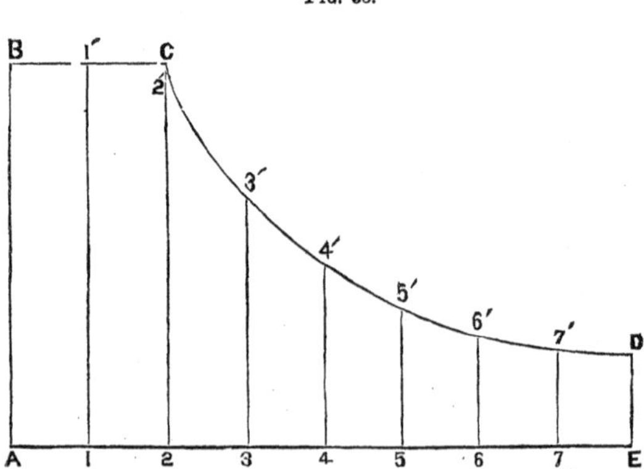

bolic curve, were there no extraneous circumstances to cause it to be otherwise; but unfortunately in practice this perfection is not attainable were Marriotte's law literally true, owing to the time required for steam to enter and leave the cylinder clearance of piston, space in nozzles between the valves, leakage of valves, piston condensation in the cylinder, &c. Fig. 33 is intended to show a perfect diagram, having all the corners well defined and the expansion line a true hyperbolic curve. From this figure we purpose explaining the manner of laying out a true hyberbolic curve. Let A E be the true vacuum line, and B C the steam line. Divide A E into any number of equal parts, and erect the perpendiculars A B, 1 1′, 2 2′, &c. Now we see that the steam follows the distance A, 2, or two divisions before it is cut off the length of the ordinate 3 3′, being three divisions from the commencement should be $\frac{2}{3}$ of 2, C′, the length of 4 4′, $\frac{2}{4}$; of 2, C; of 5 5′ $\frac{2}{5}$; of 6 6′ $\frac{2}{6}$; of 7 7′ $\frac{2}{7}$; of E D, $\frac{2}{8}$. With the lengths of all these ordinates marked on the diagram drawn through the points 3′, 4′, 5′, 6′, 7′, &c., the line C D, and you have the required curve.

An experienced engineer can tell at a glance whether an engine is in good working order from its diagram; but nevertheless, in most cases it would be well to draw the true curve, in order to ascertain how much the actual one differs from it, for by this means we can ascertain, while under way, whether the valves or piston leak; but in drawing the true curve, the clearance of the piston and space in the nozzles, &c., must be ascertained, and that much added to the length of the diagram, in order to obtain the curve accurately. Thus, supposing this space to be equal in capacity to six inches in length of the cylinder, make the diagram six inches longer than it actually is, and proceed in the manner we have shown.

Should the steam valves leak while every thing

else remains tight, the termination of the expansion line will be too high, and if the exhaust valves or piston leak, it will be too low,—allowance being made for condensation in the cylinder.

Fig. 34.

Steam.................. 10 lbs.
Rev...................... 9
Vac........... 26
Hot well...............100°
Throttle wide.

"Powhatan" stb. cylinder-top.
February 13th, 1854.

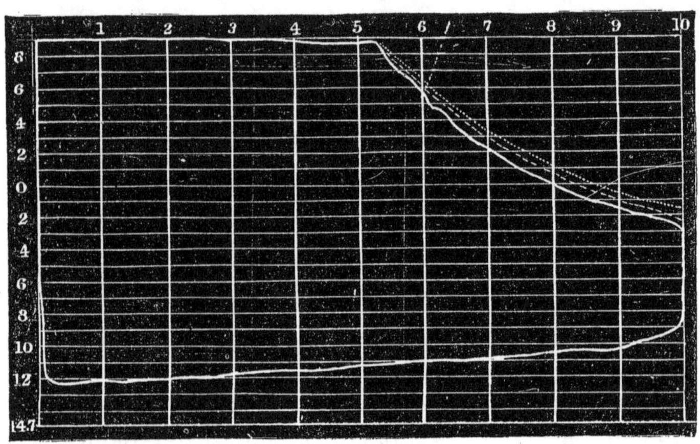

Figures 34 and 35 are two diagrams taken from the U. S. Steamer Powhatan, on the 13th of February, 1854; 34 was taken about ten minutes after 35. In both of these figures we have the true hyperbolic curves drawn in, with and without taking the clearance, &c., into account. The upper curve in small dots is the true curve, when the clearance, &c., is taken into consideration, and the lower one in large dots is the true curve without reference to the clearance, &c. In figure 35, where the steam was cut off at a very early part of the stroke, the importance of taking the clearance, &c., into consideration, is very conspicuous. The dotted lines on the right of these diagrams show the amount they are lengthened by adding the clearance, space in nozzles, &c., to them.

FIG. 35.

Steam.................. 8½ lbs.
Rev...................... 6
Vac...................... 26¼
Hot well............... 82°
Throttle wide.

" Powhatan " stb. cylinder-top.
February 13th, 1854.

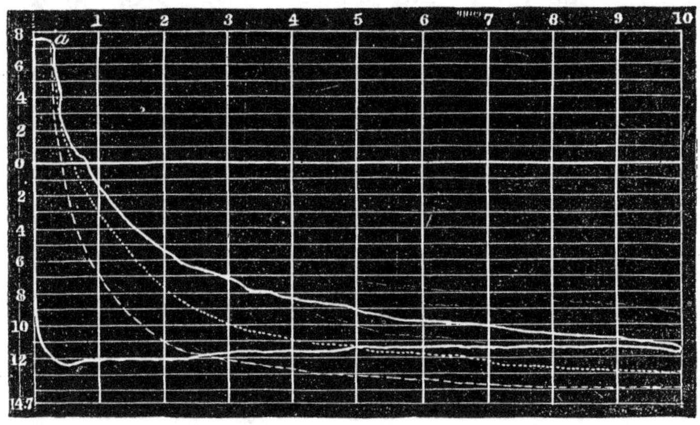

From a casual inspection of these diagrams, they seem to present an anomaly that at first is difficult to solve. Thus, in figure 34, the termination of the true expansion curve, considering clearance, &c., is about one pound *above* the actual curve, whereas in figure 35 it is two pounds below it. The first would indicate that the exhaust valves or piston leaked, and the second that the steam valves leaked, while the exhaust valves and piston were tight. Now, then, since one was taken only about ten minutes after the other, it is not at all probable that this sudden change was brought about in that short space of time; hence we must look for some defect in the engine that would occasion it. We account for it in this way: In the first case the steam valve leaked, and also the steam piston, but the piston leaked to a greater extent than the valve, that is to say, more steam passed through the piston and into the condenser from the leakage of the piston than entered the cylinder from the leakage of the valve ; therefore, the actual curve must fall below the true curve. In the second case, the steam valve also leaked, but the pressure on the piston fell so rapidly, from expansion, that it became too low to force a passage through the piston, the elasticity of the packing being sufficient, in this case—though not in the other, where it had a greater pressure to sustain—to keep it tight; hence, the true curve in this case must be below the actual curve, agreeing precisely with the conditions of the figures.

There is, however, another thing which would produce diagrams similar to those before us, and which most probably caused the formation of these, viz., leakage about the cylinder heads. Thus, supposing the stuffing box, for instance, to leak. So long as the pressure in the cylinder remained above the atmosphere, steam would blow out, occasioning the curve to fall ; on the other hand, when cutting off short, the pressure in the cylinder would soon fall below the atmosphere, and air would enter, causing the curve to rise, exactly as shown in the figures.

Fig. 36 is a diagram taken from the U. S. Steamer "San Jacinto," fitted with Allen & Wells' cut-off.

FIG. 36.

Steam in boilers......	$11\frac{1}{2}$ lbs.	November 7th, 1855, $11\frac{3}{4}$ A. M.
Revolutions...........	18	After Engine, inboard end.
Vacuum.................	$25\frac{1}{2}$	Coal 18 tons in 24 hours.
Hot well...............	104°	
Throttle 4 holes open.	Scale = $^1/_{10}$.	

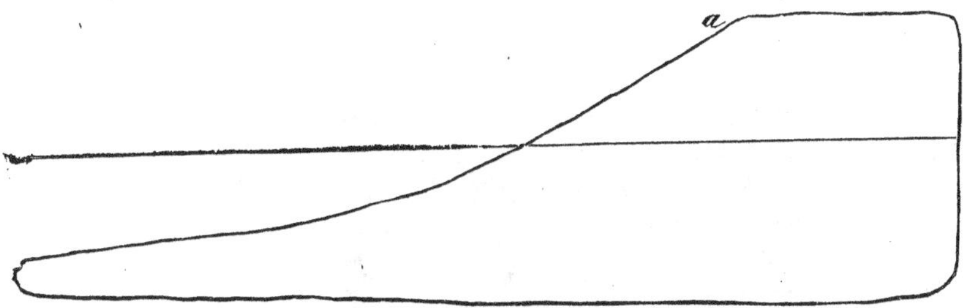

From inspection of the expansion curve of this diagram, it appears that this cut-off does not close so quickly as Sickel's, occasioning the corner *a* to be more rounding.

FIG. 37.

Steam in boilers......	9 lbs.	"Powhatan," Feb. 13th, 1854.
Revolutions...........	5	stb. cylinder bottom, working by hand.
Hot well................	100°	
Throttle.................	4	

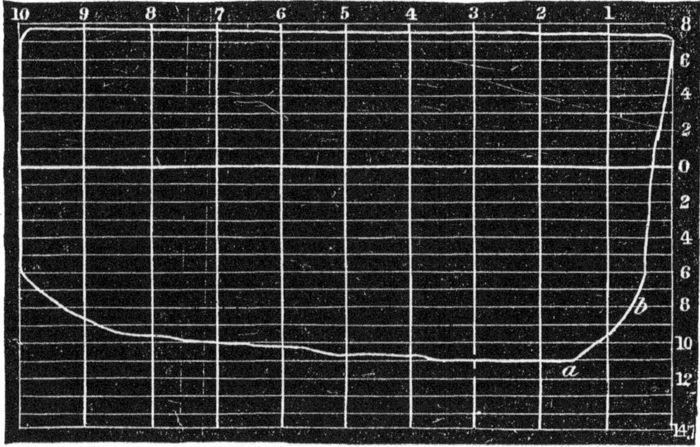

Figure 37 is a diagram showing the operation of

the valves while working by hand. This valve exhibits large cushioning and steam lead, the exhaust valve closing at *a*, and the steam valve opening at *b*, so that the engine actually passed the centre against a pressure of 6½ lbs. above the atmosphere.

Steam............	16.5 lbs.	"Powhatan" stb. air-pump, 10.50 A. M.,
Revolutions.....	9.25	January 18th, 1854.
Hot well.........	106°	

Vacuum gauge out of order.

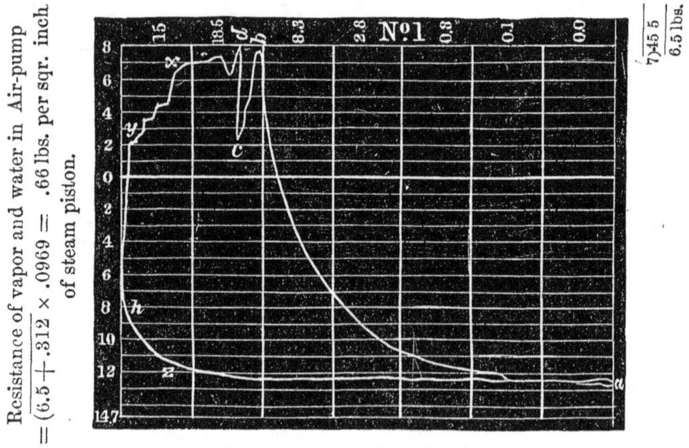

Calculated for $^1/_5$ full of water.

Number 1 is a diagram taken from the "Powhatan's" starboard air-pump. The Powhatan's air-pumps are of the lifting kind, and the piston fitted with one large brass conical valve. We will explain the diagram. At *a*, the piston being at the bottom of the stroke, starts to rise, compressing the air and vapor above it, until it arrives at *b*, at which place a sudden discharge of air and vapor seems to have taken place, and the pressure fell to *c*, from which point the pressure again gradually rose until it arrived at *d*, where the water began to be delivered and continued to the end of the stroke.

Attached to the top of the air-pumps is a pipe, running down into the bilge, for the purpose of pumping off the bilge water. Where this pipe is attached to the pump is fitted a valve, operating like an ordinary check valve, a handle being made to screw down on the top of it to keep it firmly in its seat, when there is no water in the bilge.

Steam..................	15 lbs.	"Powhatan" stb. air-pump, 10.55 A. M.
Revolutions...........	10	January 18th, 1854.
Hot well,..............	106°	
Vacuum gauge out of order.		

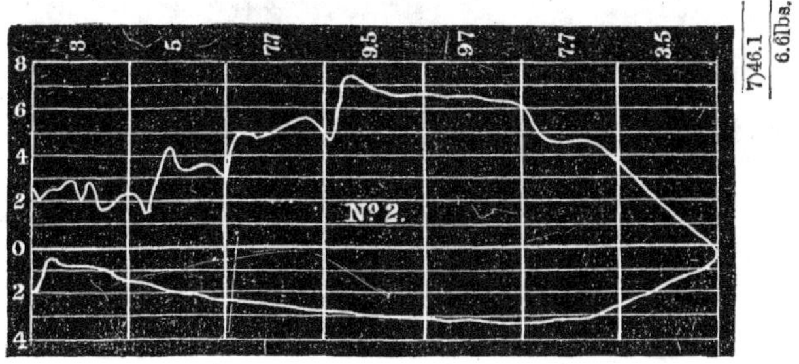

Resistance of vapor and water in Air-pump
$=\overline{(6.6+3\bar{1}2} \times .0969=)$ .6697 lb. per square inch of steam piston.
Calculated for 1/5 full of water.

There being no water in the bilge at the time No. 1 was taken, No. 2 was taken five minutes after, for the purpose of ascertaining what effect the opening of this valve and admitting air would produce. It shows that no extra power, from the admission of this air, was required to work the pump, the average pressure being about the same as in No. 1, and that the vacuum in the pumps at no time was more than 4½ lbs. There was no alteration in the vacuum, as shown by the gauge, however attached to the condenser, and the engines continued to work in the same manner as before the bilge valve attached to the air-pump was opened.

Steam....................	15 lbs.	"Powhatan" port air-pump, 11.6 A. M.
Revolutions.............	10	January 18th, 1854.
Hot well.................	108°	
Vacuum gauge out of order.		

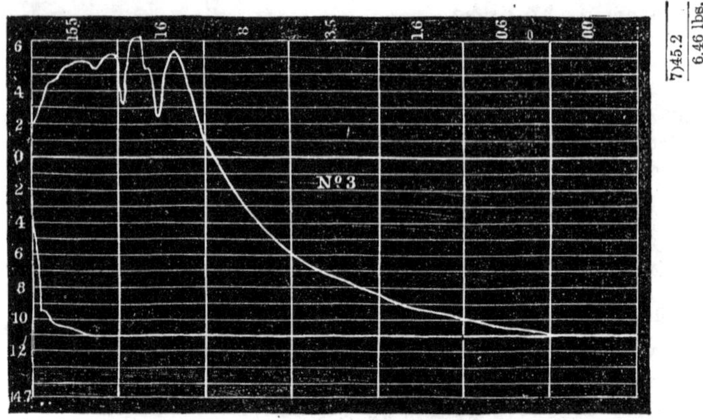

Resistance of vapor and water in Air-pump
$= (\overline{6.46 + .223} \times .0969 =)$ .6476 lb. per square inch of steam piston
Calculated for 1/7 full of water.

Nos. 3 and 4 were taken in the same manner from the port air-pump a few minutes after 1 and 2 were taken from the starboard pump.

In these diagrams the pressure at the termination of the up stroke, it will be seen, is about 2½ lbs. per square inch above the atmosphere, which is due to the height of the level of the water surrounding the ship above the top of the air-pump. The pressure increased to between 7 and 8 lbs. per square inch, as shown in other parts of the diagram, is occasioned by the friction of the water and vapor through the delivery pipes and valves. The slanting off in the diagram, No. 1, from x to y, we think partly owing to two causes: First, the decreased velocity of the piston as it approaches the end of its stroke does not expel the water with such force, and hence there is not so much fric-

Steam.....................	15 lbs.	"Powhatan" port air-pump, 11.15 A. M.
Revolutions.............	10	January 8th, 1854.
Hot well..................	108°	
Vacuum gauge out of order.		

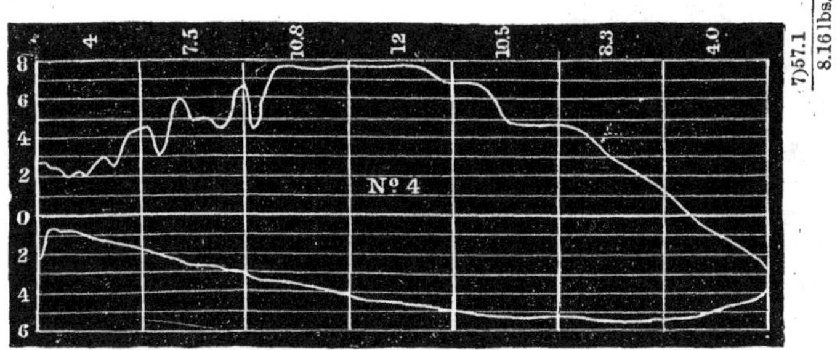

Resistance of vapor and water in air-pump
$=(\overline{8.16+223} \times .0969=)$ .8123 lbs. per square inch of steam piston.
Calculated for $^1/_7$ full of water.

tion; but this would not occasion the slanting off from *h* to *z* on the return stroke; and secondly, therefore, we are inclined to think that the string slipped or stretched a little from *x* to *y*, and recoiled again to its original place from *h* to *z*.

We will now proceed to ascertain the

Power required to work the Air-pump.

Ascertain the capacities of the steam cylinder and air-pump, by multiplying the areas of their cross-sections by the lengths of their strokes, and divide the latter by the former, which will give the ratio of the cylinder capacity to that of the air-pump. But the air-pump makes but *one* delivery stroke to every *two* strokes of the steam piston, consequently divide this ratio by two, which will give the coefficient for our present calculation, and this coefficient multiplied by the mean pressure per square inch of air-pump piston—which can be ascertained from an indicator diagram—will give the mean pressure per square inch required to expel the air and vapor.

This of course must be augmented by the weight of the water raised.

The indicator diagram will show very nearly at what part of the stroke the pump begins to deliver the water, and therefore what fraction of the pump is filled, from which can be easily ascertained the number of cubic feet of water lifted; and this number multiplied by 64.3 or 62.5, as the vessel may be running in salt or fresh water, will give the number of pounds. And the number of pounds of water lifted, divided by the area of the air-pump piston, and multiplied by the coefficient before obtained, will give the pressure per square inch of steam piston required to expel the water from the pump.

The sum of these results will give the pressure per square inch of steam piston required to work the air-pump independent of friction, an amount that is usually estimated.

Example: The capacity of the "Powhatan's" cylinder, *i. e.*, the space displaced by the steam piston per stroke, is 267.25 cubic feet: ditto in air-pump, 51.8 cubic feet; proportion of steam piston displacement to that of *half* of air-pump piston displacement, 1.000 to .0969; area of air-pump piston, 2134 square inches. The pump was filled $\frac{1}{5}$ full of water, as shown by diagram No. 1, and the mean pressure throughout the stroke was 6.5 lbs. per square inch; hence, $6.5 \times .0969 = .6298$ lb. per square inch of steam piston resistance from vapor in air-pump, and $.312 \times .0969 = .0302$ lb.

per square inch of steam piston resistance from the weight of water lifted; total = (.6298 + .0302 =) .66 pounds per square inch of steam piston, required to work the air-pump, independent of friction.

Now, supposing the mean unbalanced pressure on the steam piston per square inch to have been 20 lbs., we have 20 : .66 : : 100 : 3.3 per cent. of the total power of the engine required to work the air-pump.

CHAPTER III.

THE HYDROMETER.

THE Hydrometer is an instrument used for the purpose of determining the specific gravities of liquids. When applied to the water of marine boilers, it indicates the amount of saline matter the water contains. Figure 42 shows the kind of hydrometer usually used on board American steamers. The lower globe is filled with shot, or other weighty substance, for the purpose of keeping the instrument upright. When the hydrometer is placed in fresh water, the point O stands even with the surface of the water; when placed in water containing one pound of saline matter in thirty-two pounds of water, it stands at $\frac{1}{32}$; when the water contains two pounds of saline matter in thirty two pounds of water it stands at $\frac{2}{32}$, and so on. So that by placing this instrument in a small quantity of water, drawn from the boilers at intervals, it will show the exact density, by which we know how to regulate the blowing-off.

FIG. 42.

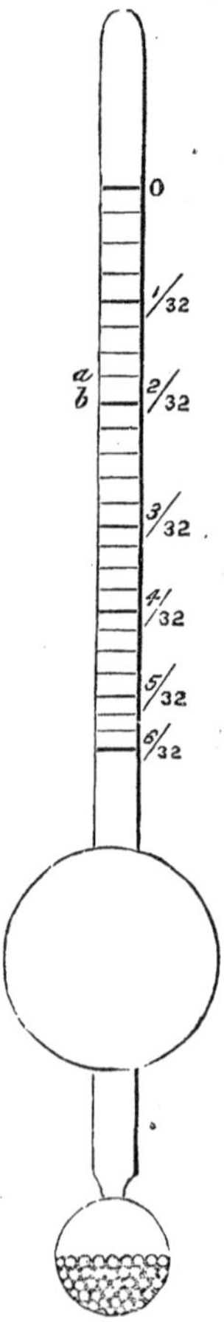

In the boilers of sea-going vessels the water is usually carried from $1\frac{3}{4}$ to 2 per hydrometer, *i. e.*, from the point *a* to *b*, figure 42. In the Gulf of Mexico, however, in the vicinity of the Florida reefs, where the water is impregnated with an unusual amount of lime, it is found not to be prudent to carry it beyond 1½.

The hydrometer, when made for a certain temperature, is not adapted to any other, but the water should be allowed to cool down to the temperature marked on the hydrometer before observing the indication, and for this purpose it becomes necessary also to use a thermometer. The hydrometers used in this country are usually graduated for a temperature of 200° Fahr. We can allow, however, for a few degrees either above or below this figure, without appreciable error—a difference of 10° in temperature making a difference of about an eighth of 1/32 in the scale. Thus, supposing the water to be at a temperature of 210°, and the hydrometer graduated for 200° to stand at *a*, or 1¾, the actual density of the water will not be 1¾, but 1⅞, or half way between *a* and *b*. On the other hand, if the temperature be 190°, and the hydrometer stand at 1¾, the true density will be 1⅝. Nevertheless, in practice, it is always best to allow the water to cool to the temperature for which the hydrometer is graduated, whenever it can be done without the waste of too much time.

It will be observed that the divisions on the scale are not of equal lengths. Thus: the distance from O to 1/32 is greater than from 1/32 to 2/32, and from 1/32 to 2/32, greater than from 2/32 to 3/32, and so on. The reason of this can be explained in this manner: When the instrument stands at O, the two bulbs, and all the tube below O, of course, are immersed, having the weight due to the length of the tube only above O to support. When it rises to 1/32 it has more weight to support, from the fact of there being more tube out of water, and it also has less bulk immersed; at 2/32 it has still more weight to support, while there is still less of the instrument immersed, and so on down to the bottom of the scale, occasioning the lengths of the divisions to become less and less.

The proportional quantity of saline matter contained in sea water, at different localities, varies very considerably, as will be seen in the following

TABLE:

Baltic Sea, ...	$\frac{1}{152}$	Mediterranean, . . .	$\frac{1}{25}$
Black Sea, ...	$\frac{1}{46}$	Atlantic at Equator, . .	$\frac{1}{25}$
Arctic Sea, ...	$\frac{1}{35}$	South Atlantic, . . .	$\frac{1}{24}$
Irish Sea, ...	$\frac{1}{30}$	North Atlantic, . . .	$\frac{1}{23}$
British Channel, . .	$\frac{1}{28}$	Dead Sea, ...	$\frac{11}{28}$

LOSS BY BLOWING OFF.

When water contains 3 per cent. by weight of saline matter, no deposit takes place at the boiling point;—under atmospheric pressure or 212° Fahr. When it contains 10 per cent. it makes a deposit of lime, principally sulphate, and at 29, 5 per cent. common salt.

The precise saturation, however, at which deposit commences to take place is not well established, but there is one thing which is well known, and that is, the higher the temperature of the water, the greater will be the deposit, and from this we conclude that common sea water would deposit a portion of its saline matter if heated to a sufficiently high temperature. The reason of the increase in deposit, as the temperature is increased, is probably owing to the expansion of the water, or the separation, as it were, of the particles.

Water carried at a density that would cause no

deposit at a temperature of 220°, would make considerable deposit at a temperature of 260° or 270°; and this is the reason why we are limited to comparatively low steam in boilers using sea water. Independent of the saving of the loss by blowing off, repairs to boilers, labor of cleaning them, &c., this is a powerful reason why inventive genius should endeavor to bring forth a reliable fresh water condenser, and why steamship owners and others, having it within their power, should encourage all such attempts, from the fact of the great advantage to be derived from carrying high pressure steam, and using the expansive principle to its fullest extent.

To the minds of those who cannot clearly see that an increase of temperature occasions an increase in deposit, a practical demonstration can be obtained by examining the crown sheets, and other parts of marine boilers, subject to the highest temperature, where it will be found the largest deposit takes place.

The deposit of lime, or "scale," as it is technically termed, on the heating surface of boilers, being nearly a non-conductor of caloric, prevents a large portion of the heat from entering the water, allowing it to escape up the chimney, and is therefore lost; and, if the deposit of scale be large, the metal of the boiler, being no longer protected by the water, becomes over-heated and "burnt." To prevent these results, a portion of the water is extracted periodically, or continuously, by the brine pump, or is discharged by the blow-off, in order to keep the density of the water below the point at which any serious deposit may take place. But as all the water discharged from the boiler has first to be heated, and as it is replaced by water of a lower temperature, a loss of heat (which is virtually a loss of fuel) is occasioned thereby. This is technically termed "loss by blowing off," and we shall proceed to illustrate the manner of calculating it. Take an example.

Supposing the density of the water entering the boiler to be $\frac{1}{32}$, and that of the boiler to be maintained at $\frac{2}{32}$, there will be one part converted into steam, and one part blown out. Supposing also the temperature of the water entering the boiler to be 100° Fahr., and the temperature of the water in the boiler to be 248° Fahr., we have all the data required.

Referring to Regnault's experiments, (page 9), we see that the total heat in steam having 248° for the sensible heat, is 1189.58°, now then

1189.58° = total heat;
100.00° = temperature of the water entering the boiler;

1089.58° = heat required from the fuel for the water to be evaporated;
248° = temperature of the water in the boiler;
100° = " " " entering "

148° = heat lost by blowing off.

Therefore, since one part (requiring 1089.58°) is converted into steam, and the other part (requiring 148°) is blown off, the total heat required of the fuel is (1089.58° + 148° =) 1237.58°; and as 148° of this is blown off, we have 1237.58 : 148 : : 100 : 11.95 + per cent. loss by blowing at the above density and temperature..

If the water had been carried at a density of $1\frac{3}{4}$ per hydrometer, one part would have been blown off as before, but only three-quarters part would have been converted into steam, hence we would have proceeded thus—

$$\begin{array}{r} 1189.58° \\ 100.00° \\ \hline 1089.58° \\ .75° \\ \hline 817.1850° \end{array}$$
= heat required from the fuel for the water to be evaporated.

$$\begin{array}{r} 248° \\ 100° \\ \hline 148° \end{array}$$
= heat lost by blowing off.

Therefore (817.185 + 148 =) 965.185 : 148 : : 100 : 15.33 + per cent.

And had the water been carried at a density of 3, *i. e.* $\frac{3}{32}$, two parts would have been used for steam and one part blown off, hence the following:

$$\begin{array}{r} 1189.58° \\ 100.00° \\ \hline 1089.58° \\ 2° \\ \hline 2179.16° \end{array}$$
= heat required from the fuel for the water to be evaporated.

$$\begin{array}{r} 248° \\ 100° \\ \hline 148° \end{array}$$
= heat lost by blowing off.

Therefore (2179.16° + 148°) = 2327.16° : 148 : : 100 : 6.35 per cent., and so on for any density. These per cents. are the losses in fuel, combustible, minus that lost from radiation and heated gases passing up the chimney.

The above calculations apply only to cases where the water enters the boiler at a density of $\frac{1}{32}$; should it enter at a lower density, the loss will be less, or a greater density more, because to retain the water in the boiler at the density assumed in the above examples, there would either have to be a less or greater quantity blown off than we have considered to be the case.

In order not to lose entirely all the heat in the water blown off, some boilers are fitted with heaters, or as they are sometimes termed incorrectly, "refrigerators." These are a series of pipes surrounded by the feed water, and through which the water leaving the boilers has to pass; by this means the temperature of the feed water is considerably increased before it enters the boiler. The following will illustrate

THE GAIN BY PUMPING WATER INTO THE BOILER AT AN INCREASED TEMPERATURE.

For this purpose two examples will be sufficient, and we will commence with the first one given above in the calculation on the loss by blowing off; viz. steam, 248°; feed water, 100°; and density, $\frac{2}{32}$. Now suppose by the application of the heater, the feed-water, instead of entering the boiler at 100°, is made to enter at 150°, what will be the saving in fuel by its application?

Solution.

1189.58° = total heat in the steam;
100.00° = temperature of the feed water;

1089.58° = heat required from the fuel to evaporate one part of water;
248° = temperature of the water blown off;
100° = " " feed water;

148° = heat lost by blowing off;

and 1089.58° + 148° = 1237.58 = total heat required from the fuel where the water is pumped into the boiler at 100°. Let us now see what the total heat will be when the water is pumped in at 150°, and the difference between these results will be, of course, the saving

1189.58° = total heat in the steam;
150.00° = temperature of the feed water;

1039.58° = heat required from the fuel to evaporate one part of water;
248° = temperature of the water blown off;
150° = " " feed water;

98° = heat lost by blowing off;

and 1039.58° + 98° = 1137.58° = total heat required from the fuel when the water is pumped into the boiler at 150°. Therefore

1237.58°
1137.58°

100° = saving in degrees;

whence 1237.58° : 100° :: 100 : 8.08 per cent. That is to say, if without the heater the boilers consumed 100 tons of coal per day, with it they would produce the same quantity of steam with 91.92 tons.

Example 2d.

Suppose that the density of the water in Example 1 was $1\frac{3}{4}$, and all the other conditions to remain unaltered, what would be the saving in that case?

Solution.

1189.58° = total heat in the steam;
100.00° = temperature of the feed water;

1089.58° = heat required from the fuel to evaporate one part of water;
.75° = part of water evaporated;

817.185° = heat required from the fuel for the water that is evaporated;
248° = temperature of the water blown off;
100° = " " feed water;

148° = heat lost by blowing off;
817.185° + 148° = 965.185° = total heat required from the fuel when the water is pumped into the boiler at 100°.

1189.58° = total heat in the steam;
150.00° = temperature of the feed water;

1039.58° = heat required from the fuel to evaporate one part of water;
.75° = part of water evaporated;

779.685° = heat required from the fuel for the water that is evaporated;

248° = temperature of the water blown off;
150° = " " feed water;

98° = heat lost by blowing off;

779.685° + 98° = 877.685° = total heat required from the fuel when the water is pumped into the boiler at 150°. Therefore

965.185°
877.685°

87.5° = saving in degrees.

Whence 965.185° : 87.5° : : 100 : 9.06 per cent. And in this manner the calculation can be made for any density and temperature.

In making calculations on the theoretical saving from the use of the heater, we have seen some engineers who calculate the loss by blowing off without it, and again with it, and take the difference between these two results for the saving; but it will require but little reflection for any one at all conversant with such subjects, to perceive the error of this mode of calculation, as it takes no cognizance whatever of the extra heat given to that portion of the water which is evaporated. The mode of calculation given above is the only correct one, as it takes into consideration all the elements.

INJECTION WATER.

After the steam has performed its duty in the cylinder, and been exhausted into the condenser, a certain amount of cold water is admitted into that vessel for the purpose of condensing it, and this quantity depends upon the temperatures of the water and the steam. We will take an example.

Suppose the temperature of the injection water to be 60°; steam as it enters the condenser, 212°; and water in the condenser, 110°. Required the proportion of injection water to the water evaporated in the boiler:

Solution.

1178.6° = total heat in the steam at the sensible temperature of 212°;
110.0° = temperature of the water after condensation;
1068.6° = heat to be destroyed;
110° = temperature of the water after condensation;
60° = temperature of the injection water,
50° difference.

Now then we see that we have 1068.6° of heat to be destroyed, and only 50° to do it with, therefore we must make up this difference in quantity; hence 1068.6° ÷ 50° = 21.372 times the evaporated water to be admitted into the condenser to condense the steam and retain the condenser at the temperature of 110°.

EVAPORATION.

Among the important elements to be ascertained in the performance of the steam engine, is the quantity of water evaporated in the boilers per unit of coal, or other fuel. In sea boilers using salt water, one pound of coal evaporates from 4 to 9 pounds of water, dependent upon the quality of the coal, the construction and cleanliness of the boilers. Those boilers are of course the best which evaporate the largest quantity, and hence the importance of knowing the exact performance of each boiler, as well as of the different kinds of fuels used in the same. To secure this desirable end we proceed thus:

Ascertain from indicator diagrams the fraction of the cylinder filled at each stroke, from which, knowing the diameter of the cylinder, we ascertain the number of cubic feet of steam required to fill that space, and to this we add the space in nozzles, clearances, &c., which gives the number of cubic feet of steam used per stroke; and the number of cubic feet of steam used per stroke, multiplied into the number of strokes per hour, and divided by the relative volumes of steam and water, at the pressure the steam is admitted into the cylinder, gives the number of cubic feet of water evaporated per hour, and the number of cubic feet of water evaporated per hour, multiplied by 64.3, (the weight in pounds avoirdupois of one cubic foot of sea water,) and divided by the number of pounds of coal used per hour, gives the number of pounds of water evaporated per pound of coal, *provided* there is no blowing off done; but wherever there is blowing off, this last result has to be increased to the extent of the loss by blowing.

Suppose for instance, proceeding in the manner given above, we find 6 lbs. of water to be evaporated per pound of coal; and the loss by blowing off to keep the water at the proper density to be 15 per cent., the remaining 85 per cent. is that which evaporates the 6 lbs.; hence 85 : 6 :: 100 : 7.06 lbs. of water evaporated per pound of coal.

EXAMPLE.—Suppose you have a cylinder 70 inches diameter by 10 feet stroke; the initial pressure of steam in the cylinder 24.5 lbs. per square inch, inclusive of the atmosphere, cut off at $\frac{1}{4}$ from commencement of stroke; clearance, &c., 10 cubic feet; revolutions, 15 per minute; coal consumed per hour, 1,500 lbs.; water carried at $1\frac{3}{4}$ per hydrometer; temperature of feed water, 107° Fahr.; required the number of pounds of water evaporated per pound of coal:

Solution.

$\frac{70^2 \times .7854}{144} \times \frac{10}{4} + 10 = 76.8125$ cubic feet of steam used per stroke; and $76.8125 \times 15 \times 2 \times 60 = 138262.5$ cubic feet of steam used per hour.

The relative volumes of steam and water at the pressure of 24.5 lbs. are 1064 to 1; hence $\frac{138262.5}{1064} \times 64.3 \div 1500 = 5.57$ lbs. of water per pound of coal, neglecting the loss by blowing off; but, according to the conditions of the example, the loss by blowing off is found to be 14.1 per cent., the remaining 85.9 per cent. is that therefore which evaporated the 5.57 lbs. of water; hence the true evaporation is found to be 85.9 : 5.57 :: 100 : 6.48 lbs. of water per pound of coal.

The above calculation takes no cognizance of the leakage of the valves, loss by radiation, or condensation in the cylinder, pipes, &c.; hence the results show too small, but it is the only standard of comparison.

Some parties calculate the evaporative power of boilers by measuring the quantity of water pumped into them during any given time, and also the quantity of coal consumed in the furnaces during the same time, and dividing the weight of the former by the latter, which they conceive gives the weight of water evaporated per unit of coal. Upon first sight this mode of operating appears very simple and correct; but unfortunately, notwithstanding its simplicity, the results are never accurate, the evaporation being always shown too large, for the very simple reason, that all the water pumped into a steam boiler is never evaporated. All boilers, and pipes, and cocks attached thereto, leak more or less, and sometimes boilers foam, occasioning water to be worked into the cylinders, and as, according to this mode of calculation, all water escaping by this means is supposed to be evaporated, the result manifestly cannot be correct.

Steam and vacuum Gauges.

As applied to the marine steam engine, the mercurial steam and vacuum gauges are the most common, though of late years there have come into use a variety of metallic gauges, many of which, from the little attention they require, appear to be very well adapted to the purpose for which they were intended.

The most prominent of these are "Schaffer's," "Hearson's," "Schmidt's," "Ashcroft's," "Eastman's," "Stubblefield's," and "Allen's." In the first three, the spring is a thin corrugated plate, upon which the steam acts, communicating motion to a hand or pointer which moves around a circular disc marked in pounds: the spring in Ashcroft's gauge is a bent tube, which the elasticity of the steam tends to straighten. Eastman's gauge is a combination of springs and levers. As these gauges are all constructed on the same principle, viz., the elasticity of metal, we shall not stop here to describe them, as it is more directly our object to deal with principles, rather than mechanical arrangements, which are the chief peculiarities of these gauges. We will pass on to the mercurial closed top vacuum gauge.

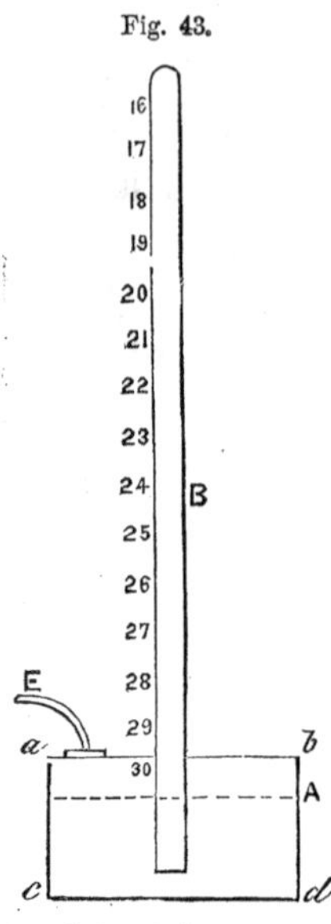

Fig. 43.

a b c d, figure 43, is a basin filled with mercury up to the point A; the tube B is also filled with mercury. The pipe *e* communicates with the condenser, and when that vessel is filled with air of the atmospheric pressure, the surface of the mercury in the basin is pressed with a pressure of about 15 lbs. per square inch, causing the tube B to remain filled; but when a partial vacuum is created in the condenser, the mercury having no longer the atmospheric pressure to sustain, falls in the tube B, and the figures marked on the scale will exhibit the extent of the vacuum. With this arrangement, therefore, there is no necessity of making the tube 30 inches in length, as all engines are supposed to maintain at least 17 or 18 inches of vacuum, and a tube long enough to show this is all that is required. Could the surface of the mercury remain constantly at A, the divisions on the scale would be of equal lengths, and one inch apart, but as the mercury rises a little in the reservoir as it falls in the tube, the lengths of these divisions vary a little, dependent upon the relative volumes of the tube and reservoir.

The aperture in the lower end of the tube is made very small, to prevent the oscillation of the mercury. At A is a small hole fitted with a screw. This is left open, while filling the gauge, as an overflow to the surplus mercury, it being so situated that the contents of the tube B is just sufficient to fill the reservoir to the point 30, or the true vacuum line.

The pressure of the atmosphere, as it varies from time to time, does not alter the indications of this gauge, inasmuch as it always exhibits the difference between the vacuum in the condenser and a perfect vacuum.

Had the top of the tube B communicated with the condenser, and the basin *a b c d* been open to the atmosphere, the gauge would then have been what is termed an open-top vacuum gauge, and would require to have been 30 inches in length—the scale being reversed, the lowest figure commencing at the bottom. With such a gauge, all variation in the pressure of the atmosphere affects its indications.

FIG. 44.

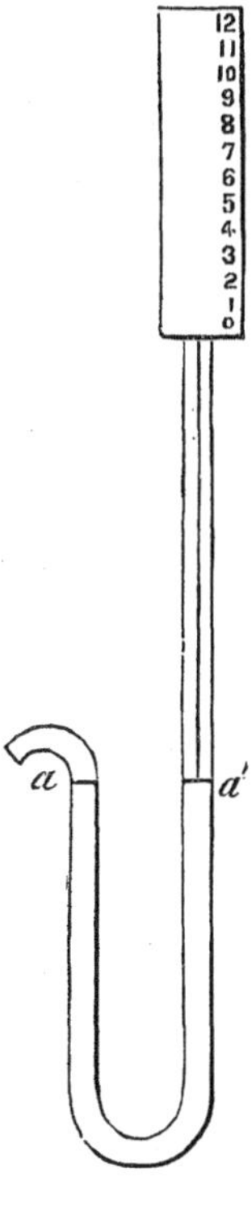

12 11 10 9 8 7 6 5 4 3 2 1 0

a *a'*

Figure 44 is a siphon steam gauge, filled with mercury to the level *a a*. The short leg connects to the boiler, and the long leg is open to the atmosphere. The steam pressing upon the mercury at *a*, forces up the stick resting on the mercury in the other leg at *a'*, showing the pressure in pounds per square inch, marked on the scale at the top of the gauge. These divisions are one inch apart, and indicate pounds pressure, for the reason that the descent of one inch in the short leg causes a rise of one inch in the long leg, making a difference in the level of the mercury of two inches, which corresponds to one pound pressure; that is to say, a column of mercury two inches high, and having a base equal in area to one square inch, will weigh in round numbers one pound.

In making a gauge, it matters not what may be the diameter of the tube, but whatever it may be, it should be uniform throughout, in order that the indications may be correct.

The stick that is put in the long leg, when there is no steam on, has one end resting on the mercury, while the other stands at 0. This stick should be made of some very light wood—soft white pine answers the purpose very well, with the lower end a little the largest, in order to have a good bearing on the mercury.

To convert this gauge into a vacuum gauge, it would be necessary only to connect the long leg to the condenser, and attach a scale to the short leg with the lowest number commencing at the top.

CHAPTER IV.

CASUALTIES, ETC.

How to act if the Eccentric be broken in an irreparable manner.

IF there be two paddle engines connected at an angle of 90°, connect the starting bar of the deranged engine, by means of a line and guide pulleys, to the cross-tail, air-pump beam, air-pump cross-head, or other part having motion coincident with the piston of the other engine, to give the bar motion in one direction, and attach a heavy weight to it, with a line running over a pulley, to give it motion in the opposite direction.

If there be but one engine, connect by similar means, to the connecting rod of the deranged engine, which will give the proper motion.

How to act when a Steamer springs aleak and commences to fill rapidly.

Put on immediately all bilge injections and bilge pumps, and shut off all other injections. If they do not keep the water down, break the joints on the bottom or side injections, and allow them to draw water from the bilge, taking care to station a man at each one to prevent any thing from passing in that would choke the valves.

Vessels are sometimes saved from foundering by

covering the leak with a sail-cloth passed over the bows and under the bottom.

If the leak be a large one, such as one occasioned by a collision, it may be possible to force a mattress, or something of that nature, into it from the outside.

How to proceed when all the feed is on and the water does not rise in the boilers.

It sometimes happens that when all the feed is on, and the feed pumps are apparently performing their usual duty, the water does not rise in the boilers, but either retains its level at the time the feed was put on, or gradually falls. In this event, one of two things must be manifest—either that the water does not enter the boiler, or if it does enter, is escaping through some other orifice. The first thing, therefore, to do, is to examine the check valve to see if it is in operation. This can be done by applying the ear to the chamber, to ascertain if the valve rises and falls, at each stroke of the pump, and also by applying the hand to the pipe, immediately below the check valve, in order to ascertain if it be cool. If these are found to be all right, examine the blow-off cocks, and all other water connections with the boilers, to ascertain if they be closed; some of which, in all probability, will be partially open, but if they should all be found closed, the pump must be pumping air into the boilers instead of water. The next step would therefore be, to examine the pump and induction pipe, in order to ascertain and stop the air leak.

Upon examining the check valve, should it not be found in operation, the next step would be to examine the pump, to see if it was hot; also relief and pump valves, to see if they were gagged; and lastly, the eduction pipe, to see if it were burst—either of which causes would prevent the pump from delivering water. A feed pump may get hot from four causes:

First. There may be so small a quantity of injection water used as to cause it, when delivered to the hot well, to be of sufficiently high temperature to heat the pump.

Second. Friction, occasioned from muddy water, or tight packing.

Third. The check and delivery valves may be caught up or very leaky, allowing the hot water from the boiler to run back to the pump.

Fourth. External application of heat, the pump being situated near the boiler or other hot body.

A feed-pump cannot deliver water when hot, for the reason that the vapor constantly generated within it, by its elasticity prevents the induction valve from opening and admitting water.

Should the feed pipe burst, it can be repaired temporarily by wrapping it with canvas coated with white lead; this being secured by strong twine or marline, wound closely around the pipe the full length of the canvas.

Should the pipe be split open for a considerable distance, it might first be closed with wood or iron clamps, as came most convenient, before applying the canvas and twine.

Foaming.

Foaming, or priming, as it is sometimes termed, is violent ebullition or agitation of the water, occasioned by an undue relation of temperature between the steam and water. Thus, supposing a large quantity of steam to be suddenly taken from the boiler, the pressure of steam is immediately reduced below what is due to the temperature of the water, and the result is a sudden rising up of the water from all parts of the boiler. Foam can, therefore, be defined to be a mixture of steam and water. Boilers are known to be foaming when the water does not come out of the gauge cocks solid, or when there is a considerable agitation of the water in the glass gauges.

To suppress foaming, put on a strong feed and blow off, cut off shorter or partially close the throttle. Oil or melted tallow, injected into the boilers through the feed pumps, will also prevent foaming, but these are somewhat expensive expedients.

Boilers constructed with insufficient steam room, are most likely to foam, because at each stroke of the piston a large proportion of the steam is taken from the boiler, and the pressure therefore becomes materially reduced. Boilers also constructed in such a manner as to prevent the easy escape of steam from the surfaces on which it is generated, are likely to foam. Thus, supposing there be a large amount of heating surface on the crowns and other parts towards the bottom of the boiler, and that the steam generated on these surfaces in consequence of coming in contact with the flues, tubes, braces, &c., can find but a comparatively small exit to the surface of the water, the result will be, that where it does escape, it will force a large body of water up, mixing it with the steam.

To carry too much water in boilers will cause them to foam by reducing the steam room. Running from salt to fresh water, or *vice versa*, will also cause foaming; in the former case, because fresh water boils at a lower temperature, but a satisfactory explanation of the latter case appears to be difficult to arrive at. The boilers of sea steamers, when running in muddy rivers, usually foam considerably.

It sometimes occurs, while the boilers are foaming badly, that the engines have to be stopped in order to take soundings, or from other causes. Now, the first thing after stopping the engines, in any case, is always to try the water; for it will mostly always be found to be lower when the engines are standing still than when under way, but when the boilers are foaming, it is of the highest importance to try immediately the height of the water, for as the foaming ceases after the engines are stopped, it may happen that the water has fallen entirely out of the gauges and left the flues, in which event, if the engines were going to be started again in three or four minutes, the better plan would be to open the safety valve to keep the water foaming, so as to keep the flues covered, and when the engines are started again to put all the feed on. But if the engines were going to stand still for a considerable time, blow off a portion of the steam, if it be too high, dampen the fires a little, and put on the auxiliary feed.

The Condenser heats.

When engines are standing still, it sometimes occurs that the condenser gets so hot, that when it becomes necessary to start again, the pressure has become so great in it, that the injection water will not enter. Leaky steam and exhaust valves will alone cause this, but in no case should it ever be allowed to occur. When an engine begins to get hot, the cracking noise in the condenser, and about the foot valves, will always indicate what is going on, time enough to check it, which can be done by giving a little injection, and causing the engines to make two or three revolutions back and forth. If, however, the engine should become too hot to take the injection water, the only plan will be to blow through, or pump water into the condenser if there be such an arrangement, or to cool the condenser by external application of cold water.

If when under way it is indicated by the gauge that the engine is gradually losing its vacuum, apply the hand to the condenser, in order to ascertain if it be getting hot, and if such be found to be the case give a little more injection; but if that does not help the cause, give more still. If the vacuum continues to grow less, the probability is that the injection pipe has become choked; in which event shut off that injection and put on another. Should both the bottom and side become choked, inject from the bilge. Should the bilge injection also be out of order, the engine will have to be stopped, and the snifting valve secured down (if there be one) while the injections are blown through to clear them. Sea weed, and things of that nature, sometimes get over the strainers of injection pipes, preventing the entrance of water.

Most if not all marine engines of modern construction are fitted with a thermometer to the hot well, to ascertain the temperature of the water, which is usually carried from 100° to 115° Fahr. This instrument is very important, in order to maintain an even temperature (the sense of touch of the engineer's hand not being delicate enough for that purpose), for it may often occur that there may start small leaks about the condenser and exhaust pipe joints, which would cause a decrease in the vacuum, and, as without the thermometer, the first impulse would be to give more injection, with it we would turn our attention to finding and stopping the leak. This can be done by holding a lighted candle around the joints, and wherever there is a leak the flame will be drawn in. To stop it, mix a little putty, of white and red lead, and apply it to the crevice; the presence of the atmosphere will force it in.

Getting under way.

When lying in port, where the steam will not be required for at least four or five days, it is proper that the water should be blown or pumped out of the boilers, and a portion of the man and hand-hole plates removed, to allow a circulation of air. When, therefore, the order is given to get up steam, the first thing is to see that all these plates are put on, and the joints properly made, and this duty should receive the direct superintendence of the engineer having charge of the same; for should any one of them leak badly after the steam is raised, the departure of the ship might be delayed some hours in consequence. After this duty has been properly attended to, open the blow-off cocks and run the water up in the boilers to the proper level, or, if the boilers are so situated that the water will not run up high enough, finish the supply with the hand pumps, wood the furnaces while the water is entering the boiler, and when the proper height of water is attained start the fires. If it be important to raise steam quickly, start the fires as soon as water is discovered in the gauges, continuing the supply while the fires are burning. As a small quantity of finely split wood, with a little shavings or oily waste placed in the mouth of the furnaces, is all that is necessary to start the fires, the back part of the furnaces, particularly in boilers with inferior draft, should be covered with a layer of coal to keep out the cold air.

In raising steam it has been the custom to recommend that the valves of the engine be blocked open, so as to allow the heated air from the boilers to pass in and warm up the engine before steam begins to be generated; but as in many cases this is attended with considerable trouble, and as the advantages to be derived from it are very small, it hardly appears to the author's mind to "pay." The safety or vacuum valve should, however, be kept open until steam begins to form, in order to let the heated air escape. The strain upon boilers being from the inside, they are constructed and braced with the special view of withstanding this strain, many of the braces being entirely useless in sustaining a pressure from without; marine boilers are therefore fitted with a small valve opening inwards, and weighted so as to open and admit air whenever the pressure from within falls to about five pounds per square inch below the atmosphere. These valves are called differently by different parties, as follows: *vacuum valve*, *air valve*, reverse valve, &c.

After steam has been raised to 3 or 4 lbs., the engine should then be blown through and warmed up, and after sufficient steam is raised to move the piston, the engine should be turned over two or three times, to see that every thing is right, before reporting ready.

On Coming into Port.

After the engines are no longer needed, before hauling the fires, after a long run, it would be well to try the pistons and valves, in order to ascertain if they be leaky. To try the piston, open the water valve on one end of the cylinder and the steam valve on the opposite end; if the piston leaks, the steam will escape through the water valve. To ascertain if the steam valves leak, open the water valves on both ends of the cylinder. To ascertain if the exhaust valves leak, open the steam valves and any cock in the exhaust side of the steam chest or exhaust pipes.

While under way it may be discovered that there is a slight thump in the engine when passing one or or the other or both centres, and the indicator having been applied shows the usual lead, the inference is that some part of the working engine is loose; it is important, therefore, to find out what it is on coming into port. To do this place the engine on the centre, and give the piston steam suddenly by raising and lowering the starting bar; observe closely the cross-head, crank-pin, main-shaft, and other main connections, to see where the jar is. Should it not be discovered after this, jam the cross-head fast, so as to prevent the slightest motion, and then give steam as before, in which event, if the thump be still felt, the piston will doubtless be found to have worked a little loose.

If it be the intention to remain in port several days, before hauling the fires, sufficient steam should be raised, if the boilers be capable of bearing the pressure, to blow all the water out of the boilers. After the boilers become cool, the hand-hole plates, over the furnaces particularly, should be taken off, to examine the crowns, where the greater amount of scale will be found deposited, and from which we can judge if the boilers require scaling. Mere dampness in boilers is found to be injurious, by occasioning a rapid oxidation, and in order to prevent this, one or two hand-hold plates should be taken off the bottom of the boilers, in order to let the water drain out dry. It would be well also to remove a man-hole plate from the top of the boilers to allow a circulation of air. If these things cannot be done it will be better to keep the boilers filled with water, rather than a small quantity in the bottoms. In damp climates, such as the Isthmus of Panama, light fires should be made in the ash-pits occasionally.

Scaling Boilers.

Notwithstanding the water in the boilers is not allowed to exceed in density $1\frac{3}{4}$ to 2 per saline hydrometer, it will be found after a time that a quantity of scale, composed principally of lime, has accumulated on the crown sheets, tubes or flues, and other parts of the boiler. If this be allowed to remain the metal will become overheated and burned; it becomes necessary, therefore, to remove it, which can be alone done by mechanical means. Sharp-faced "scaling hammers" can be used to knock the scale off those places that are within the arm's reach, and long bars flattened at both ends, and sharpened, called "scaling bars," will knock it off the more remote parts. In the Martin tubular boiler, which is accessible in every part, it is only necessary to condense the steam in the boilers for a day or so after the ship comes to anchor; this will soften the scale so that a gang of men may be put into them as soon as the man-hole plates are removed, and scrape off all of it in a few hours. The scale, however, must never be allowed to exceed the thickness of writing paper.

It has been proposed in some quarters to heat the tubes or flues by burning shavings, or some other such substance in them, and then to cool them off suddenly by pumping cold water upon them, the sudden contraction causing the scale to crack off. This plan, however, to our mind, does not deserve much favor, and never should be resorted to, if the scale can be reached in any other manner, for the production of leaks will mostly always be the result.

It is, however, hoped that engineers will soon be relieved from this duty, and steamer owners benefited by the introduction of fresh water condensers into all sea steamers.

Preparatory to coming to Anchor, or securing to the Wharf.

Fifteen or twenty minutes before coming to anchor, or making fast to the wharf, the chief engineer should be informed of the fact by the officer of the deck, or some other person informed on the matter, so that the fires can be allowed to burn down, and the pressure of steam permitted to fall to such an extent that the necessity for blowing off is avoided. By this means the great nuisance of blowing off steam is not only obviated, but there is a considerable saving in fuel, the fires being permitted to burn down sufficiently low to supply only the amount of steam required while working the engines by hand, rendering it much easier also on the firemen (whose duties on any occasion are arduous enough) by having a very light, instead of a very heavy fire to haul.

In coming to anchor it is usually well to pump a little extra water into the boiler, so as to insure a proper supply while operating the engines by hand.

When it is desired to raise steam, the order from the captain should always be what time it is intended to get underway, leaving to the discretion of the chief engineer to start the fires at such time as he may consider proper, in order to secure steam and every thing ready at the proper time.

Regarding the Fires while under Way.

Small as this may appear in the eyes of one not practically conversant with the management of the steam engine, it is one of the most important things that the engineer is called upon to regulate: on the one hand, that a proper and uniform supply of steam is maintained, and on the other, that more fuel is not consumed than is actually necessary to produce the result. Different fuels and differently constructed boilers require the fires to be regulated in a different manner, and notwithstanding the repeated efforts, the adoption of specific rules, which shall apply alike to all, is positively absurd. A few general hints, however, touching the leading features, may be useful to those who have not had much experience in this matter, but they must bear in mind, nevertheless, that actual service and observation for themselves, will alone make them proficient, no matter how well they may understand the chemistry of coal, or the natural laws governing the combustion of matter.

The proper supply of atmospheric air, and the proper time for the combustion, are the important elements in the consumption of coal. A slow rate of combustion, and a moderate draft, always producing a better evaporative result, than when the fires are urged, occasioning them to be more rapid; and hence, on no occasion, should "blowers" be resorted to, if the proper supply of steam can be maintained without them.

The fire should be spread uniformly all over the grate bars, and in the use of bituminous coal, should be from 6 to 8 inches in thickness, but with anthracite coal, 4 or 5 inches will be thick enough. So long as the ash pit remains bright, there is no necessity for slicing or stirring up the fire, but whenever the spaces between the bars become choked with clinker, or ashes, it will be indicated by the darkness in the ash pit, and, if burning bituminous coal, a slice bar should be run in through the stoke holes or furnace doors to break up the fire and clear out the air spaces. A pick applied from below is also very useful in this respect. In the use of anthracite coal the pick alone should be used; the breaking up of the surface of such fire,—as it does not amalgamate or run together, forming a crust like the bituminous,—prevents the regular uniform combustion by allowing too much air to enter among the disturbed parts of the coal, it requiring considerable time for them again to unite in regular ignition after being once disturbed. It is very important that no part of the grate bars be left bare, as the admission of cold air, through such space, deadens the fire, and cools the flues. It has been ascertained of late, that better results are obtained by admitting air through a number of small holes in the furnace doors, on the plan of W. Wye Williams, Esq., of England.

No two furnaces should be fired at the same time; the fresh coal of the one should be fairly ignited before a new supply is added to another, in order to keep a regular supply of steam. Anthracite coal requires less frequent firing than bituminous, but with either, the coal should not be thrown upon any particular part of the furnace, but uniformly all over it. Before firing with bituminous coal, it is well to break up the upper crust of the fire, which sometimes amalgamates so closely as to exclude the proper supply of air. The trouble with most firemen is, that they are disposed to heap their fires too much, particularly in front, sometimes half way to the crowns; this they do for three reasons: first, because they suppose the larger the fire the greater the supply of steam; second, the more coal there is piled in at one time, the less frequent they will have to fire; and third, it requires much less labor to shovel the coal into the mouth of the furnace, than to supply it uniformly, all over the grates. No coal larger than one's fist should be allowed to enter the furnace, nor in cleaning the fires, should more than one be cleaned at the same time, which should be done at stated intervals, unless it so happens, that they all or many of them, have got so dirty that a further supply of coal is useless, when the engine can be throttled off a little while the cleaning is going on. In cleaning anthracite fires, care should be taken not to reduce them too low, otherwise they will take a long time to recover.

In cleaning fires, as well as when supplying them, the furnace doors should not be kept open longer than necessary, admitting an undue supply of cold air; and the party, therefore, who, performing his duty as well, does it the quickest, is the best fireman.

The slower a steamer runs the greater distance she will perform with the same amount of fuel, provided she has not an adverse tide or head winds to contend with; with men-of-war, therefore, it often occurs that the saving of fuel is a more important consideration than high speed, and for this reason the consumption of coal is reduced far below what would be required to keep the vessel up to her maximum speed. This can be done in two ways: either by shutting off a portion of the furnaces entirely, by shutting the ash pit doors and closing up the cracks around them with wet ashes, or else reducing the quantity of coal consumed in each, by covering the back part of the grates with a thick layer of ashes. When the diminution in the quantity of coal is not very large, this latter plan is the better, by retaining the original heating surface at the same time that the combustion of coal is allowed to go on very slowly, an end very desirable to secure. When, however, the reduction in coal is very considerable, some of the furnaces can be shut off, while the back ends of the grates of the remainder can be kept covered with ashes. Men-of-war sometimes proceed at half or less speed, and as a large extent of boiler surface occasions considerable loss from radiation, in such cases it will be more economical to shut off some of the boilers and continue with a moderate supply of fuel in the remainder. The furnaces and ash pits of the boilers shut off should be closed tightly, to prevent cold air from passing in to cool the surfaces of the other boilers, or to injure the draft.

After a boiler is shut off, the steam should not be allowed to escape, but to remain in it and condense, to freshen the water.

Patching Boilers.

Inasmuch as all things constructed by human hands are liable to decay, steam boilers are not exempt from this infallible law; they therefore frequently require to be patched, new stay bolts and braces to be put in, old rivets cut out and replaced with new ones, &c. In patching boilers, wherever the defective part can be reached so as to work at it well, it is best to cut it out and rivet a patch on, calking the seams; but as this cannot always be done, the most common practice is to put a patch over the defective part, securing it with bolts and nuts, or tap bolts, and making the joint with stiff putty, composed of white and red lead, and a small quantity of fine iron borings. A piece of sheet lead fitted over the place to be patched, will answer for the pattern to make the patch by, which, however, before the joint is made, should be fitted snugly to its place while hot.

Owing to imperfection in the iron, small cracks are sometimes discovered in the flues or other parts of the boiler, subject to a high temperature. Should these not be more than two or three inches in length, they can be stopped by drilling holes and putting in three or four small rivets, hammering the heads well down so as to cover the crack.

A leaky stay-bolt, or rivet, has, like the toothache, but one sure remedy, and that one is to cut it out and put in a new one.

In cutting out a stay-bolt fitted with a socket, the latter can usually be saved and retained in its place, ready to receive another bolt; but sometimes a screw bolt is cut out which has to be replaced with a socket bolt, and as this may be in such part of the boiler which cannot be reached by the arm, or tongs, a very good plan to get the socket in its place, is to pass a string through both holes and secure the ends, dropping the centre down and hauling it out through a hand hole; cut the string in two, pass the ends through the socket, join them together again, and haul the socket to its place. In the fitting of sockets, it is very important that they should be the exact distance between the sheets, with the ends filed square, otherwise the sheets will be drawn out of shape.

Sweeping Flues.

One of the most disagreeable parts of the duties is that of cleaning flues, from the fact of its dirtying every thing round about or in the vicinity of the boilers, the slightest draft being sufficient to waft the light dry ashes in every direction. A little water sprinkled on them before they are hauled out of the connections or smoke-boxes will prevent this in a measure, the damper and ash-pit and furnace doors being closed, to prevent the men from being suffocated who go inside. The lower flues, particularly, are apt to leak a little, and the salt water, mixing with the ashes, forms a solid mass, which can only be removed by being cut out, the flue brush being of no avail. The hammer and chisel, and long, sharp-pointed bars, and sledge, are best adapted to the purpose. In the use of these instruments, care should be taken that they be not driven through the metal or under the seams.

Ash Pits.

The ash pits should be cleaned out every watch, and the ashes thrown overboard, picking out first any lumps of coal that may have fallen among the ashes. When not running at full speed, a portion of the cinders may be thrown upon the fires again, after damping them with a little water. So also should fine bituminous coal be dampened before being supplied to the furnaces, the arguments to the contrary notwithstanding; for though it does take a little heat from the fire to evaporate the water mixed with the coal, a saving is effected, by preventing the coal from being drawn—particularly in boilers with strong draft—through the flues and lodged in the connections, or out of the smoke-pipe. No more water, however, should be put on the coal than just sufficient to dampen it.

Smoke-pipe Stays

Require to be looked to occasionally, when made of rope, as they grow a little slack from time to time. These should always be adjusted while the pipe is hot; otherwise, if they be set up while the pipe is cool, the expansion after it becomes heated will, in all probability, "carry" either the stays themselves away, or the band securing them to the pipe. In a gale of wind, when the ship is rolling heavily, these stays should be looked to, in order to tighten any of them that may have become slack, so as to throw the strain alike on all. Hemp rope is a very inferior article for such purpose as stays for smoke pipes, and we can see no good reason, unless it be prejudice, (which is always a good reason to those under such influence,) why it has been so long retained. Good wire rope looks better, is cheaper, and will last a great deal longer, and requires much less attention.

Grate Bars, &c.

When fitted new, are usually allowed plenty of play, both fore and aft and sideways, to allow for expansion after they become heated. The spaces at the end of the bars, however, become choked up with ashes, which become, by and by, so hard as to form almost a solid mass, defeating the objects for which they were left. These spaces, therefore, in port, should be cleaned out occasionally.

Ash pits, in port, should also be well cleaned and painted, to prevent oxidation. At sea, no water should be thrown into them upon the ashes, but they should be kept as dry as possible. With these precautions, they will last as long as other parts of the boiler. Boilers unused for any considerable time should be kept dry of water, and have fires made occasionally in the ash pits, to evaporate all interior deposit of dampness—the neglect of this precaution is the sole cause of the oxidation and deterioration of all boilers when not in use.

Broken Air-Pump.

Should the air-pump become broken in an irreparable manner, and the engine be a single one, there is but one thing that can be done, and that is to work non-condensing. If there be two engines, we have three resorts: to work the broken engine non-condensing, to disconnect from the crank pin and proceed with one engine, or, if there be facilities on board, to join the exhaust of both engines with a pipe, and use one air-pump and one condenser for both engines. This latter plan was tried very successfully for a short run on board the U. S. Steam Frigate "Powhatan," on the China station, in the summer of 1855. Peculiar facilities were, however, offered in this case, as the exhaust side pipe of each engine had a man-hole in it, to which the connecting pipe was joined.

In running under such circumstances, care should be taken not to overload the air-pump.

Broken Cylinder Head.

Water may be worked over into the cylinder suddenly, from boilers foaming badly, or otherwise, faster than it can escape through the water valves, and being nearly non-compressible, something must give way, the cylinder head, or bottom, being the most likely thing to go. In such an event, if there be a spare one on board, put it on; if not, while the old one is being repaired, if it be reparable, the following plan can be resorted to; Disconnect the steam and exhaust valves from the damaged end of the cylinder, if the engine be fitted with poppet valves, and let the atmospheric pressure force the piston in one direction, the steam being used for the opposite direction. Should the engine be fitted with a slide valve, close up the opening into the damaged end of the cylinder, by fitting in, steam-tight and in a substantial manner, a block of soft wood. This should not, however, be resorted to, except in cases of great emergency. Cylinder heads should have manhole plates of less strength than the heads; this would prevent the destruction of heads in all cases.

The selection of Coal.

The kinds and qualities of coals are so varied that no general rules can be given for their selection, but there is one point, however, which we think will not be disputed, and that one is, whenever there is a choice, the only sure plan is to select the best; for, though its first cost may be a little more, it will prove to be the cheapest in the end. What economy is there in purchasing one coal because it can be obtained 10 or 15 per cent. cheaper than another, when there will be burned, to produce the same effect, from 20 to 25 per cent. more than would be burned by the better kind? Yet this is a thing of daily occurrence. But, regardless of the money view, there are other disadvantages attending the use of the inferior coal. From the fact of there being more burned, the firemen have more to supply to the furnaces, and it requires, on their part, greater care and attention to keep the fires in good order; thus imposing extra duty on a portion of the ship's crew whose energies are usually overtaxed. Besides, to convey the vessel a given distance, an extra quantity has to be taken on board, which, in the case of merchant ships, diminishes their freight capacity, or, in war ships, lumbers the deck with a useless number of bags.

Some boilers are best adapted to bituminous coals, others to anthracite, and the one or the other of these coals which should be selected, depends upon the circumstances, therefore, for which they are intended.

In the selection of coals, it is an object to obtain those free as possible from earthy impurities. Slate, and such like matter, is to be avoided. Sulphur in bituminous coals makes them the more liable to spontaneous combustion. So also receiving them on board wet will endanger spontaneous ignition. Coals which have been exposed a long while to the rays of the sun, particularly in tropical climates, undergo a gradual decay, reducing their evaporative qualities.

Safety Valve.

Steam, when once commencing to blow off, will not cease when the pressure has fallen to the pressure due to that for which the safety valve is loaded, but will continue to blow-off until the pressure has fallen some pounds below this. This is owing to the increased area which the steam has to act upon when the valve is open over what it has when the valve is closed, occasioned by the bevel of the valve face. In a heavy sea, the safety valve may be forced open for a short time, even when the pressure is below that for which the valve is loaded, by the oscillation of the ship.

CHAPTER V.

MISCELLANEOUS.

The Theory of the Paddle Wheel; the Radial compared with the Feathering Wheel.

To all those whose minds have a tendency to probe beyond the superficial crust of any thing that may be presented to their consideration, the theory of the action of the paddle wheel on the water must be one of interest, and any thing, therefore, tending to make this subject the more clear, cannot fail to receive the proper attention and a careful perusal.

In regard to the paddle wheel, many theories have been advanced, some of them so positively absurd that it is difficult for us to conceive how they ever found their way into print. Even in reference to the subject of centre of pressure of the paddles, such rules as the following have been put forth from quarters to which we should have looked for more correct information:

"The circle described by the point whose velocity equals the velocity of the ship, is called the *rolling circle*, and the resistance due to the difference of velocity of the rolling circle and the centre of pressure is that which operates in the propulsion of the vessel." * * *

Rule: "From the radius of the wheel subtract the radius of the rolling circle, to the remainder add the depth of the paddle board, and divide the fourth power of the sum by four times the depth; from the cube root of the quotient subtract the difference between the radii of the wheel and the rolling circle, and the remainder will be the distance of the centre of pressure from the upper edge of the paddle. The diameter of the rolling circle is very easily found, for we have only to divide 5280 times the number of miles per hour by 60 times the number of strokes per minute, to get an expression for the circumference of the rolling circle, or the following rule may be adopted: Divide 88 times the speed of the vessel in statute miles per hour, by 3.1416 times the number of strokes per minute; the quotient will be the diameter in feet of the rolling circle."

Now, then, I suppose no one who has given the subject the slightest attention would imagine, for one moment, that so long as the immersion remained constant, a difference in the slip of a common radial wheel would make a difference in the centre of pressure of the paddles; yet if any one will take the trouble to work out the centre of pressure of any wheel by the above rule with different slips, he will find the centre of pressure continually changing. To suppose such a thing to be true would be as absurd as to suppose the centre of pressure of a plank immersed vertically in a stream moving at the rate of 10 miles per hour, to be in a different place from what it would be should the stream move at the rate of 5 miles per hour.

We have thought it advisable, therefore, to go into this subject the more fully, and give the following as an illustration of our views:

It is generally admitted that the total loss of effect, or power, in the common radial wheel, is the sum of the losses of the oblique action on the water and the slip. The former is calculated by taking the mean of the squares of the sines of the angle of incidence at which the paddles strike the water, or which is the same thing, the means of the squares of the cosines of the angles of the arm and water; for one angle is the complement of the other. This will appear plain from an inspection of figure 1. A C is the arm, making

Fig. 1.

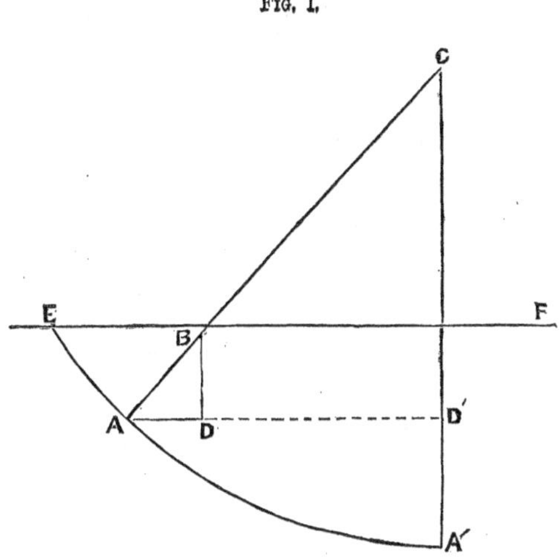

an angle at C, with the vertical line C A′; A B, the breadth of the paddles, and E F, the surface of the water. Now, it is manifest, that, inasmuch as the vessel is moving in a horizontal direction, the line B D at right angles to that direction, represents the only portion of the paddle A B that is efficient in propelling the vessel, and the line A D represents that portion of the paddle that tends to lift the vessel out of the water, which, consequently, as it produces no propulsive effect, must be entirely lost. But the line A B, being the breadth of the paddle, we will suppose represents the pressure it exerts on the water, which, according to the resolution of forces, is divided into two other pressures. A D, tending to lift the vessel, is the useless pressure, and B D, at right angles to the vessel's path, is the efficient pressure, or the portion that is utilized in propelling the vessel. Power, however, is not composed of pressure alone, but is compounded of pressure and velocity, and as the velocities of the columns of water, having A D B D for the base depend upon the lengths of those lines respectively; that is to say, if we double the length of either one of them, say B D, for instance, diminishing the angle at C, we not only double the quantity of water displaced in any given time, but it is also displaced with double the velocity; the power, therefore, developed is the product of these two, or as the square. Hence, it follows that, since A D represents the useless pressure, the square of that line must represent the useless or lost power; or, more correctly, the loss of useful effect, and the square of B D, the power that is applied to propelling the vessel. Now, then, considering A B to be unity, the square of B D will be the square of the natural sine of the angle B A D, and the square of A D the square of the natural sine of the angle A B D; but the triangles A B D, A C D′, being similar, the angles at B and C are equal, and the loss of effect is, therefore, simply represented by the square of the sine of the angle that the oblique arm makes with the perpendicular; but as the angle is continually changing, as the arm moves through the water, we have to take the mean, and the more numerous, therefore, the divisions are made, the nearer correct will be the result.

Thus, supposing, as per figure 2, a wheel 26 feet diameter, from outside to outside of paddles, 6 feet immersion of lower edge of paddles, and 20 inches breadth of paddles, the loss from oblique action is calculated as follows, the arc being divided into divi-

Fig. 2.

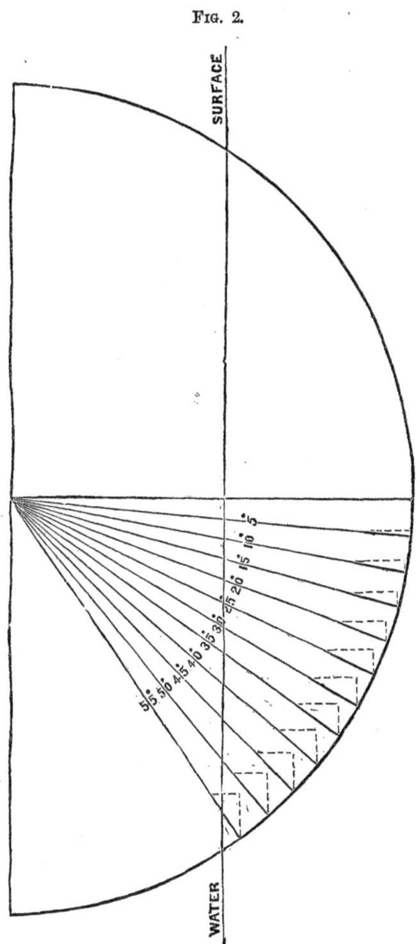

sions of 5° each, which are considered sufficiently numerous for practical purposes :

	Angles of Incidence.	Sines of the Angles of Incidence.	
½	55°	.81915	.33550 = half of the square of sine
1.	50°	.76604	.58681 = square of sine.
1.	45°	.70711	.50000 "
1.	40°	.64279	.41317 = "
1.	35°	.57358	.32899 = "
1.	30°	.50000	.25000 = "
1.	25°	.42262	.17860 = "
1.	20°	.34202	.11697 = "
1.	15°	.25882	.06698 = "
1.	10°	.17365	.03015 = "
1.	5°	.08716	.00759 = "
1.	0°	.00000	.00000 = "
1.	5°	.08716	.00759 = "
1.	10°	.17365	.03015 = "
1.	15°	.25882	.06698 = "
1.	20°	.34202	.11697 = "
1.	25°	.42262	.17860 = "
1.	30°	.50000	.25000 = "
1.	35°	.57358	.32899 = "
1.	40°	.62279	.41317 = "
1.	45°	.70711	.50000 = "
1.	50°	.76604	.58681 = "
½	55°	.81915	.33550 = half of the square of sine.
22			5.62952

As 22 : 5.62952 : : 100 : 25.588 per cent. of the power applied to the wheels.

Half of the square of the sine at the angle of 55° is taken, because the paddle in that position is only half immersed, consequently only half the power can be expended on it as if entirely immersed; and the angles are put down twice, because the loss is the same after the paddle leaves the vertical position as before it reaches it. The power in the latter case being expended in forcing the water downwards, and in the former case in lifting the water, neither of which assists in propelling the vessel, the only tendency being to lift the bow, and depress the stern.

Slip.

The loss of effect from slip is usually considered the difference between the velocity of the centre of pressure of the paddles and the velocity of the vessel.

Thus, if the velocity of the centre of pressure of the paddles exceeds the velocity of the vessel by 18 per cent. of the speed of the paddles, 18 per cent. is considered the loss of effect from slip. This we conceive to be an error. The 18 per cent. is the difference between the velocity of the paddles and the velocity of the vessel, nothing more; and, therefore, simply represents the slip in per cent. of the paddles, but not the *loss* of effect from slip. For it has been shown that the loss resulting from the oblique action of the paddles on the water, is as the squares of the sines of the angles of incidence, and if we suppose the wheel to be immersed to its axis, the loss from this cause on the paddle, when in the horizontal position—the angle being 90°—is 100 per cent., and if the loss from slip of 18 per cent. be added to that, we have a total loss of 118 per cent., or more than the power applied. A positive absurdity. Or, again, supposing the vessel to be made fast to the wharf, the difference between the velocity of the paddles and the velocity of the vessel will be 100 per cent., and as the loss from oblique action cannot, from this circumstance, be any less than if the vessel was moving ahead, there will be a total loss of the power applied to the wheels of 125.588 per cent. A result equally absurd.

At the angle of 45° it has been seen that only .70711 part of the area of the paddle is effective in propelling the vessel, and that at this angle the velocity of the column of water driven aft is only .70711 of what it is when the whole area of the paddle is effective, hence the power expended in slip $= .70711 \times .70711 = 5$, the slip in the vertical position being considered 1.

Now, then, if 18 per cent. is the loss from slip

when the paddle is in the vertical position—which must be the case if its velocity exceeds that of the vessel by 18 per cent. of its own speed—from what has just been shown, at the angle of 45°, the loss cannot be more than half of 18, or 9 per cent. The same reasoning will demonstrate, that at the angle of 30° the loss from slip cannot exceed $\frac{3}{4}$ of 18, or 13.5 per cent. Thus we see the loss from slip goes on decreasing from the vertical to the horizontal position, at which place it becomes nothing. We can, therefore, approximate very nearly to the true loss in the present radial wheel, by taking the mean of these losses at the angles as laid down in figure 2. They are as follows:

At 0°	= 18 −	.0000	= 18	per cent.
" 5°	= 18 −	.1366	= 17.8634	" "
" 10°	= 18 −	.5427	= 17.4573	" "
" 15°	= 18 −	1.2056	= 16.7944	" "
" 20°	= 18 −	2.1055	= 15.8945	" "
" 25°	= 18 −	3.2148	= 14.7852	" "
" 30°	= 18 −	4.5000	= 13.5000	" "
" 35°	= 18 −	5.9218	= 12.0782	" "
" 40°	= 18 −	7.4371	= 10.5629	" "
" 45°	= 18 −	9.0000	= 9.0000	" "
" 50°	= 18 −	10.5626	= 7.4374	" "
" 55°	= $\frac{18 - 12.078}{2}$		= 5.9220	" "
			138.3343	
			2	
Doubled for both sides of the vertical position			276.6686	
			18.0000	
			294.6686	

$\frac{294.6686}{22}$ = 13.394 per cent. of the power applied to the wheels.

The same result is obtained as follows:

100.000 (power applied) — 25.588 (oblique action) × 18 per cent. (slip of the vertical paddle) = 13.394 per cent.

We have, therefore, for a total loss in this radial wheel, 25.588 + 13.394 = 38.982 per cent. of the power applied to it.

Feathering Wheel.

Let us take a feathering wheel, of the same diameter of centre of pressure, *i. e.*, 26 feet 4 inches in diameter from outside to outside of paddles—same immersion, breadth, and number of paddles, and see how it compares with this.

It is conceived by some that the only losses in this kind of wheel are the friction of the eccentrics, &c., and the slip, but there is another loss with deep immersions, or light slips, occasioned by the *drag* of the paddles as they enter and leave the water.

In figure 3, the paddles are supposed to be vertical from the time they enter until they leave the water, and the positions of the arms will be seen at the degrees there laid down. The perpendicular lines drawn across the arcs are intended to represent the breadth of the paddles. It is plain that while the axis of the paddle moves from A to B, it moves horizontally the distance A C, and vertically the distance C B, and, supposing the vessel to be moving with the same velocity as the paddles, it will travel the distance A B, while the paddle travels horizontally the distance A C. Now, the distance A C being less than A B, the paddle in this position cannot be giving out any power, but must be keeping the vessel back, by carrying a column of water before it, the base of which is equal to the area of the paddles, and the length equal to the difference in the lengths of the two lines.

Fig. 3.

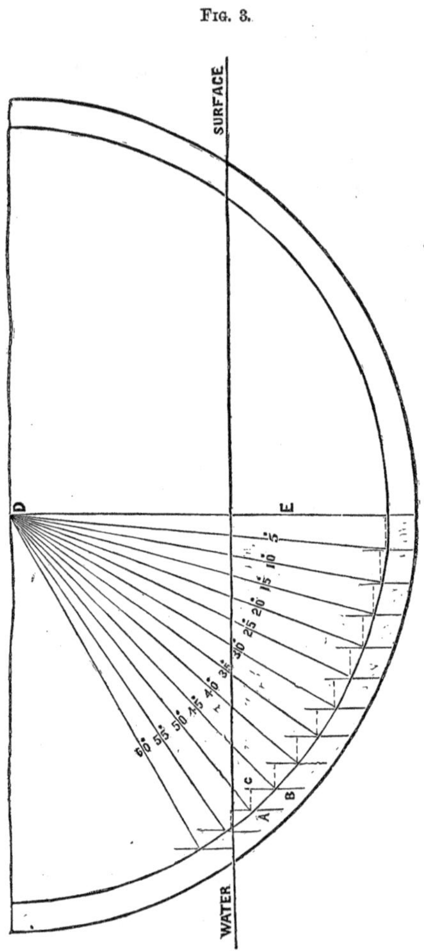

If A B be represented by unity, A C will be represented by the natural sine of the angle A B C, and if the arc be supposed to be divided into an infinite number of parts, or composed of an infinite number of straight lines, A B will be at right angles to A D, and, by consequence, the angle A B C will be equal to the angle D A E; and as the sine of B represents the distance traveled horizontally by the paddle, the sine of D A E must manifestly represent the same thing, but the sine of D A E is the cosine of D, which therefore represents the horizontal velocity of the paddle at the angle of 50°, its circular velocity being 1. The difference between these two lines is, therefore, the loss from *drag*, supposing there to be no slip, but as all paddle wheels must have *some* slip, when they are propelling a vessel, the line A B, diminished by the amount of slip, will represent the distance traveled by the vessel, and the *loss* from drag will therefore, instead of being the difference between A B and A C, be the difference between a fraction of A B and the whole of A C, dependent upon the amount of slip. If this fraction of A B be just equal to A C, the loss from drag in this position becomes 0; for, though the paddle be giving out no power to the vessel, it occasions no resistance to the vessel's progress through the water, because it is moving horizontally precisely as fast as the vessel itself; and if the fraction be less than A C, the resistance will, of course, be on the after instead of the forward side of the paddle, and it must, in consequence, necessarily be assisting in propelling the vessel.

Now, then, from the above, it must be evident to any one, that so long as the paddle, after it enters the water, is moving horizontally at a less rate than the vessel, it cannot be giving out any power, but must be an actual resistance to the vessel's progress through the water. Taking figure 3, and giving the wheel the same mean loss from slip as the radial wheel, viz.,

13.394 per cent., we will ascertain the loss from slip at the different angles there laid down, and attend to the drag afterwards, which is merely slip in the opposite direction, or what might be termed *negative* slip.

To give this wheel the same mean loss from slip as the radial wheel, it has to have on the arm when in the vertical position, or

	Cosines.			
At 0°			26.225	per cent.
" 5°	= .99619 — .73775	=	25.844	" "
" 10°	= .98481 — .73775	=	24.706	" "
" 15°	= .96593 — .73775	=	22.818	" "
" 20°	= .93969 — .73775	=	20.194	" "
" 25°	= .90631 — .73775	=	16.856	" "
" 30°	= .86603 — .73775	=	12.828	" "
" 35°	= .81915 — .73775	=	8.140	" "
" 40°	= .76604 — .73775	=	2.829	" "
" 45°	=		0.000	" "
" 50°	=		0.000	" "
" 55°	=		0.000	" "
			134.215	
			2	
Doubled for both sides of the vertical position,	- -		268.430	
			26.225	
			294.655	

$\frac{294.655}{22} = 13.394$ per cent. of the power applied to the wheel lost by slip.

At the angle of 55° the paddle is .445 part immersed, but, being so near, we have taken it at a half for simplicity, and for like reason have considered the paddle at 50° entirely immersed.

It will be seen from the above, that the paddle, from the time it enters the water until after it passes

45°, is traveling horizontally at a less rate than the vessel, and the same effect ensues as it rises out of the water; there must, therefore, be a loss from drag or negative slip. Let us see what this amounts to.

Cosines.

$$\text{At } 55^\circ = \frac{.73775 - .57358}{2} = 8.208 \text{ per cent.}$$

" $50^\circ = .73775 - .64279 = 9.496$ " "

" $45^\circ = .73775 - .70711 = 3.064$ " "

20.768

2

Doubled for entering and leaving, 41.536

$$\frac{41.536}{22} = 1.933 \text{ per cent.}$$

We have, then, for a total loss in this wheel, slip (13.394 per cent.) + drag (1.933 per cent.) = 15.327 per cent. of the power applied to it.

The total loss in the radial wheel having been shown to be 38.982 per cent. (and in the feathering wheel 15.327 per cent.), we have 23.655 per cent. in favor of the feathering wheel. But of the whole power applied to the engines, about 20 per cent. is expended in overcoming friction of ditto, friction of load on working journals, working air and feed pumps with their loads, &c. Consequently, only 80 per cent. reaches the wheels, and 23.655 per cent. of 80 per cent. equals 18.924 per cent. of the total power applied to the engines in favor of the feathering wheel.

To stand off against this, we have the friction of the eccentrics, &c. (an amount that, perhaps, can only be estimated) extra weight and wear and tear of the wheels.

It will be seen also from the above, that the difference between the velocity of the feathering wheel and the vessel being 26.235 per cent. of the speed of the wheel, and the difference between the velocity of the radial wheel and the vessel being 18 per cent. of its speed, it follows that, making the same number of revolutions, the speeds of the vessels will be as 73.775 to 82, or as 1.00 to 1.11; consequently, the speed of the feathering wheel will have to exceed the speed of the radial wheel 11 per cent. to give the vessel the same velocity, but this speed of the wheel is as shown—consequent upon there being less resistance to the paddles—attained by an expenditure of 18.924 per cent. less power.

Centre of Pressure.

The centre of pressure of a rectangular plane immersed in a fluid, the upper extremity of which is even with the surface of the fluid, is $\frac{1}{3}$ from the bottom; but, inasmuch as the pressure is as the depth, when its upper extremity is below the surface of the fluid, this law no longer holds good. To ascertain the centre of pressure in such case, "Jamieson on Fluids" gives the following practical rule deduced from elaborate mathematical calculations:

"Divide the difference of the cubes of the extremities of the given plane below the surface of the fluid, by the difference of their squares, and two-thirds of the quotient will give the distance of the centre of pressure below the surface, from which subtract the depth of the upper extremity, and the remainder will show the point in the centre line of the plane in which the centre of pressure is situated."

This rule can be applied directly to the feathering wheel, by taking the mean immersion of the paddles as they move through the water, and assuming figure 3 to be of the same diameter from outside to outside of paddles, as figure 2, viz: 26 feet, we find the mean immersion of the lower edges of the paddles, after their upper extremity gets below the surface, to be

$$\frac{(29.23+37.84+45.59+52.44+58.32+63.09+67.02+69.78+71.44)}{19}$$

$2+72=55.87$ inches, and upper edge 35.87 inches.

The mean centre of pressure of the paddles in these positions is $\left(\frac{55.87^3-35.87^3}{55.87^2-35.87^2}\right)\frac{2}{3}=46.59-35.87=10.72$ inches from top, or 9.28 inches from bottom, and the mean centre of pressure from the time the paddle enters until it leaves the water,

$$\frac{9.28\times19+(6.62+3.25)2}{23}=8.52 \text{ inches from the bottom.}$$

In the radial wheel, however, as the outer extremity of the paddle moves more rapidly than the inner extremity, and as the resistance is as the square of the velocity, the centre of pressure must be considerably nearer the outer extremity on this account. One-third from the bottom, in this case, is, therefore, probably, not much out of the truth; but as a portion of the paddle only part of the time is immersed, we take the mean of the third of that portion and a third of the whole breadth of the paddle during the time it is entirely immersed.

Thus: $\frac{(20\div3)\times21+(10\div3)2}{23}=6.37$ inches from the bottom, showing the centre of pressure under these circumstances to be $(8.52-6.37=)\,2.15$ inches nearer the lower edge of the paddle in the radial than it is in the feathering wheel.

Practical Remarks on the Foregoing.

From what has been shown, it would appear that the use of the feathering wheel over the radial wheel, from the great saving it effects, would lead to its universal adoption; but, unfortunately, the practical difficulties are such that its use is confined within very narrow limits. The increased weight of the wheel, occasioned by the eccentrics, levers, arms, &c., required to work the paddles, amounting, in some cases, to several tons, causing the pillow-block brasses to wear away very rapidly, is a sad objection, to say nothing of the excessive friction they produce. Besides, the pins operating as the axis about which the paddles vibrate are found to wear away very rapidly, requiring not only to be replaced frequently, but the noise and jar occasioned from the wear becomes very objectionable. The latter objection, however, can be removed by the use of lignumvitæ pin bearings.

The Screw Propeller.

The great advantages derivable from the successful adaptation of the screw propeller, particularly to vessels of war, became well understood in its early history, and inventive genius set to work thenceforth to perfect this important invention; all kinds of propellers sprang into use, many of them possessing neither the merit of novelty nor usefulness. One, two, three, four, five, six-bladed, true screws, expanding pitch and no screw at all, are among the number that have been tried experimentally and practically since the introduction of the screw propeller, and, strange as it may appear, notwithstanding the large share of attention it has received, the theory of the screw propeller is yet not generally understood; but, to our mind, this is owing to one great cause; and that is, to the very important fact, that those who have undertaken to explain and illustrate it, have apparently thought it more important to give the history and accounts of the experiments—though both very useful in themselves—than to explain the leading features and the laws governing its action. Besides, a practical engineer does not wish, or if he did, has not the time to spare, to examine large volumes to find what might be condensed into a few pages. We have, therefore, determined to make our remarks on this subject brief, and to confine them to those points which we think are the more important, allowing the student to build upon them for himself.

The surface of a screw blade may be supposed to be generated by a line revolving around a cylinder, at right angles to the axis, at the same time that it moves along it, and should the revolving motion be a constant ratio to the motion lengthwise, it will be a true screw. Should such a screw as this, Fig. 4, be developed upon a plane it will form a right-angled triangle, in which A B is the pitch, A C the circumference described by the extremity of the blade, and B C the line described by any point in the periphery of the blade by one convolution of the thread. To make this the more clear, suppose the triangle A B C to be wound round a cylinder, having a circumference equal to A C, and suppose at C we start to trace a line around the cylinder,

Fig. 4.

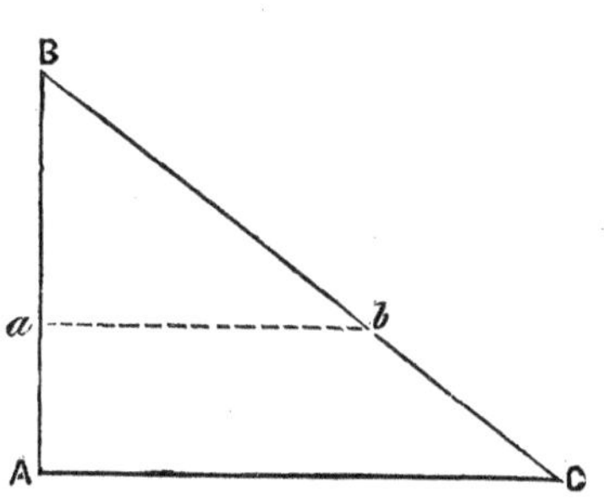

moving along it at the same time in a constant ratio, and that when we have gone all the way around, arriving over the starting point C, (C and A will be one and the same point in the case supposed) we have reached the point B, C B will be the line described, which is technically termed the *directrix*, and A B, being the distance moved in the direction of the axis, will be the pitch. Should the line A B be a curve, instead of a straight line, the screw would have an increasing or expanding pitch, instead of an uniform pitch. Figure 5 will illustrate this: Let the curve B C be the curve of the blade, and the dotted lines B *b*, C *c* be tangents drawn to this curve, it will be seen that, at different points in the curve B C, the velocity of rotation remaining constant, the velocity lengthwise of the axis A B varies, growing greater as we approach B. This is what is termed an expanding pitch; that is to say, the pitch at the anterior portion of the blade, is less than the pitch at the posterior portion. The object of such a pitch is this: the anterior portion of the blade striking upon water at rest, encounters the resistance due to a solid body moving through water at rest, but this portion of the blade puts the water in motion, it being a yielding medium, so that when the posterior portion of the blade follows it has to act on water in motion, instead of water at rest, and in order, therefore, to make the resistance due to all parts of the blade alike, the pitch of the posterior portion of the blade is increased to the extent of the motion given to the water by the anterior portion.

Fig. 5.

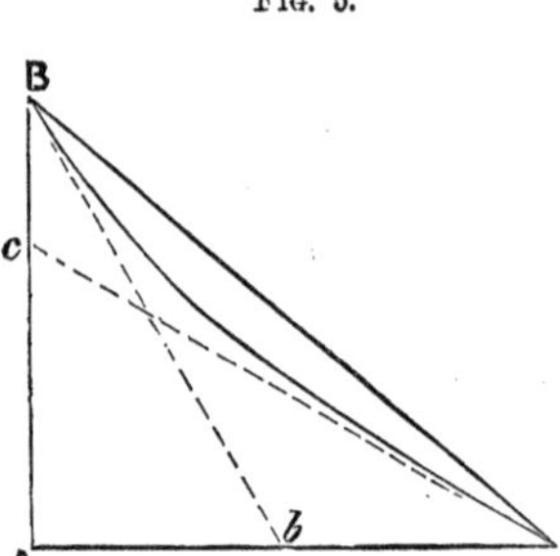

To measure the pitch of a screw blade, did it extend all the way round the shaft to a full convolution of the thread, all we would have to do, would be to measure along the line of the shaft from any point in the blade to any point directly over it, and the distance would be the pitch, or the distance traveled in the direction of the axis by one convolution of the thread; but since in practice, in order to secure the proper resisting area, a full convolution of the thread is not required—a very small fraction of it being used—it becomes necessary, therefore, to find the pitch from this fraction. Taking figure 4, for instance, let B *b* be the length of the blade, measured on the periphery, and A C the circumference described by the extremity of the blade, B *b* will be the fraction of the blade used, and B *a* the fraction of the pitch. We know, therefore, that, starting from B, and traveling along the line B *b*, when we arrive at the point *b*, we have traveled along the axis the distance B *b*, and from this we can ascertain what distance will be moved along the axis by continuing all the way round until we arrive at C, which will be the pitch. Practically, we can measure this in two ways: measure the length B *b* of the blade, and also B *a*, the length in line with the axis, we have then two legs of a right-angled triangle, from which we ascertain the third, *a b*. Now, then, knowing the circumference described by the extremity of the blade, we derive the following simple proportion:

As *a b* : the whole circumference : : B *a* : the whole pitch.

Or we proceed thus: Lay a straight-edge across the face of the propeller, at right angles to the axis, and a bevel on the periphery of the blade, and look them out of wind, the angle enclosed by the two legs of the bevel will be the angle B *b* *a*, which is termed the "angle of the propeller;" and hence, if B *b* be supposed unity, the fraction of the pitch of the one blade will be (B *a*) the natural sine of the angle B *b* *a*, therefore, knowing the angle B *b* *a*, and the length of the blade B *b*, we ascertain the pitch thus:

As cosine *b* : whole circumference of propeller : : sine *b* : to whole pitch.

The pitch can also be determined by construction, without any calculation whatever. Thus, supposing the line *a* *b* represents the whole circumference of the propeller, we draw the line B *b* at the angle to *a* *b* ascertained from measurement, and erect the perpendicular *a* B, which will give the pitch required.

In a true screw, it matters not whether we take the angle at the periphery or any other part of the blade; for, though the angle will be different, increasing as we approach the centre, the pitch will be the same, it only being necessary to know the circumference at the point where we measure the angle.

Should the blade not be a true screw, but an expanding pitch, we have to take the angle at two or more points, by drawing tangents to the curve, and take the mean, for the mean angle of the blade. Thus, in figure 5, the mean of the angles B *b* A and *c* C A will give the mean angle of the blade.

Some propellers are made to expand from hub to periphery, instead of from anterior to posterior portion of the blade.

To ascertain the pitch of such a propeller, take the mean of the angles at several points in the blade, and proceed as above. In order to ascertain the pitch of any propeller, it is always proper to take the angles at two or more points in the blade, from which we learn whether it expands from hub to periphery, whether it be true screw, or no screw at all.

The fraction of the pitch, as we have explained it above, is the fraction of the pitch of one blade, but as screw propellers usually have two, three, four, six, &c., blades, constituting fractions of a double-threaded, treble-threaded, four-threaded, six-threaded, &c., screw, the sum of these constitute the fraction of what is usually termed the fraction of the pitch of the screw; that is to say, if the screw have three blades, and the fraction of the pitch of one of those blades be $\frac{1}{12}$, the real fraction of the pitch will be 3 times $\frac{1}{12}$, or $\frac{1}{4}$; for it evidently matters not, as far as this is concerned, whether the screw be in one, or divided into a dozen parts.

How to lay down a Propeller.

Knowing the diameter, number of blades, and fraction of pitch, we intend to use, we proceed thus:

FIG. 6.

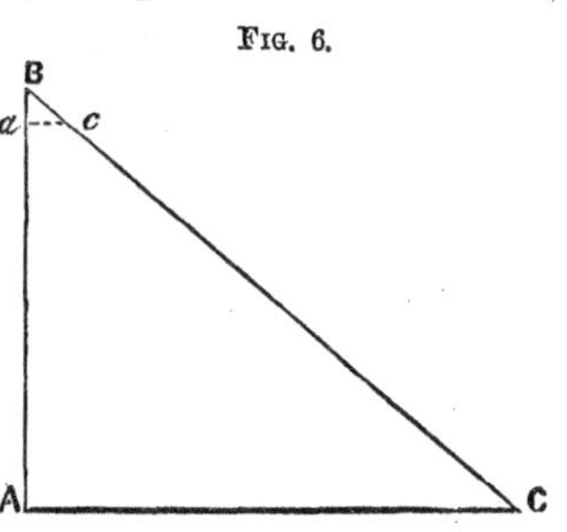

Taking figure 6, for instance, draw the line A C, equal the circumference of the extremities of the blades, and from A erect the perpendicular A B, equal the pitch; join B C. Now, then, supposing we desire the propeller to have four blades, and the fraction of the pitch to be $\frac{1}{3}$, lay off B *a*, equal to $\frac{1}{12}$ B A, and draw *a c*, parallel to A C. *a c* will be the circumference of the extremity of one blade viewed as a disc. Then, taking figure 7, we describe the circle *a b c*, equal A C, figure

6, and also the smaller circle, equal the circumference of the hub of the propeller; divide the larger circle into four equal parts, and, from the centres thus obtained lay off *a d*, *h i*, *b f*, *e c*, each equal to *a c*, figure 6, and draw lines from each of these points to the centre, terminating in the hub; such will be the projection of a four-bladed, true screw propeller, viewed from the stern, from which the longitudinal elevation can be drawn. The dimensions of the sections of the blade depend upon the diameter of the propeller, the material of which it is constructed, and the pressure it has to sustain.

Fig. 7.

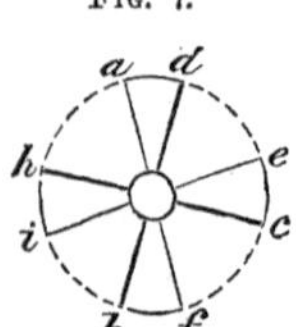

Centre of Pressure.

All solid bodies moving through a fluid have a certain point called the centre of pressure, which is the point where the outer and inner pressures just exactly balance. In a screw propeller, the radius of the circle, which is equal to half the area of the whole circle, described by the periphery of the blades, is the centre of pressure from centre of motion. Thus, if a propeller be 16 feet diameter, the area of the circle described by the extremity of the blades $= 201.06$ square feet, and the radius of the circle, having an area equal to half this, is 5 feet $7\frac{7}{8}$ inches, consequently the centre of pressure in this propeller is 5 feet $7\frac{7}{8}$ inches from the centre of shaft.

The centre of pressure can also be ascertained in the following manner:

$$\frac{1+4+9+16+25+36+49+64}{1+2+3+4+5+6+7+8} = \frac{204}{36} = 5 \text{ feet } 8$$

inches, nearly as before.

The line per sketch represents the radius of the propeller, and is divided into divisions of 1 foot each; the more numerous, of course, the divisions are made, the nearer correct will be the result.

In these calculations, the area of the hub is neglected.

The above rule holds good so long as there is no variation in the pitch from hub to periphery; but should the pitch vary in this direction, the velocity of the column of water driven aft from different parts of the blade will also vary, effecting the centre of pressure correspondingly.

Slip.—The slip of a screw propeller is the difference between the velocity of the propeller and the velocity of the ship.

EXAMPLE.—A propeller having 20 feet pitch makes 70 revolutions per minute, which propels the vessel at the rate of 12 knots an hour, required the slip, the sea knot containing $6082\frac{2}{3}$ feet?

ANSWER.

$20 \times 70 \times 60 = 84000 =$ speed of propeller in ft. per hour.
$6082\frac{2}{3} \times 12 = 72992 =$ " vessel " "
$\overline{11008} =$ slip in feet.

84000 : 11008 : : 100 : 13.1 = slip in per cent. of the speed of the propeller.

Thrust.—A propeller being put in revolution throws a column of water off from the blades in line with the axis of the propeller, which, as explained above, is the slip; the resistance of this water acting upon the propeller blades, tends to force the shaft inboard, which resistance has to be sustained by heavy bearings called *thrust bearings*, and the amount of this resisting pressure is called the *thrust*. In order, in practice, to ascertain the extent of the thrust, an instrument called the *dynamometer* is attached to some part of the shaft. This instrument consists of a combination of levers or weighing beams, to the final end of which is attached a spring balance, or scale, which indicates the pressure in pounds; and this pressure being augmented by the number of times the levers are multiplied, gives the total pressure, or thrust on the shaft. And the total thrust being multiplied into the distance moved over in a unit of time by the vessel, shows the actual power absorbed in propelling the vessel.

In the application of the dynamometer, care must be taken that it receives the entire thrust of the shaft before the indication of the scale is noted.

Did the propeller and steam piston travel through the same distance in any given time, and were all the power applied to the piston transmitted to the water through the propeller, the total pressure upon the steam piston and the thrust of the propeller would be identical, but since such is never the case, we ascertain the theoretical thrust, thus:

$$\frac{\text{Total effective pressure on piston in lbs.} \times 2 \text{ length of stroke in ft.} \times \text{No. of revols. per min.}}{\text{Pitch of propeller in feet} \times \text{number of revolutions per minute.}}$$

= Theoretical thrust in lbs. The difference between this and the actual thrust, shows the amount lost in friction of engines, propeller, and load, overcoming resistance to edge of propeller blades, working pumps, etc. The loss from slip is independent of this.

Strain upon a Screw Propeller-blade.

We can best illustrate this by an example.

Given, circumference of centre of pressure of a 3 bladed propeller, 30.9 feet; distance from hub to centre pressure 41 inches; pitch 22.5 feet; thrust 12700 pounds: required, the strain upon each blade at the hub.

SOLUTION.

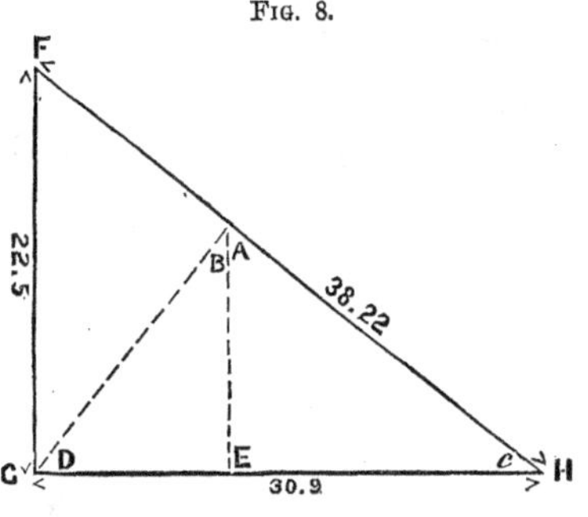

FIG. 8.

Let F G H be the development of the helix on a plane, draw B D at right angles to F H, and A E at right angles to G H. Trigonometrically, we ascertain the angles at A and D to be each = 37° 9′, and at C and B to be each = 52° 5′, and the lengths of the lines A E, B D, to be relatively as 1.000 to 1.237.

Now, inasmuch as the whole thrust can be supposed to be concentrated in the centre of pressure of the blade, and as the 12700 lbs. is in a line with the axis, it follows that, if the line A E represents the direction and amount of this thrust, the line B D, at right angles to the propeller blade at the centre of pressure, according to the resolution of forces, will represent the resultant of the pressures on the blade, or the total pressure tending to break it. But inasmuch as there are three blades, the pressure will be divided equally among them all; therefore, each has to sustain but a third of this pressure; hence

$$\frac{12700 \times 1.237 \text{ (proportion B D bears to A E)}}{3} = 5236$$

lbs. pressure on each blade at the centre of pressure.

The pressure at the hub on each blade equals 5236 lbs. × 41 ins. = 214676 lbs. acting with the leverage of one inch.

EXAMPLE 2D.—Suppose, in example 1, the breadth of the blades at the hub to be 32 inches, and the propeller to be made of composition, capable of sustaining a pressure per square inch of cross-section of 520 lbs., acting through the leverage of 1 inch; required, the mean thickness of the blade at the hub?

SOLUTION.—The strength of beams is directly as their breadths and the squares of their depths, and inversely as their lengths. In the example before us, the propeller resolves itself into a simple beam; we have, then, $\frac{214676 \times 1}{32 \times 520} = 12.9$ inches = square of the thickness, and $\sqrt{12.9} = 3.59$ inches in thickness.

Helicoidal Area.—As has already been shown, the development of the helix on a plane is the hypothenuse of a right-angled triangle, having the pitch of the screw for the height; and the circumference, corresponding to the radii of the helix, for the base. Now, as the propeller can be supposed to have an infinite number of helices, each one becoming longer and longer as we approach the periphery, which alter the lengths at the same time, of the hypothenuse and base of the triangle, we will suppose the propeller to be divided into a number of concentric rings, taking the centre line of each, for the helix or hypothenuse of the triangle; the circumference corresponding to radii of said helix for the base, and the pitch for the height, from which we have all the elements required for the calculation.

To make this the more clear, take the triangle B A C; the lines B 1, B 2, B 3, B C, represent the

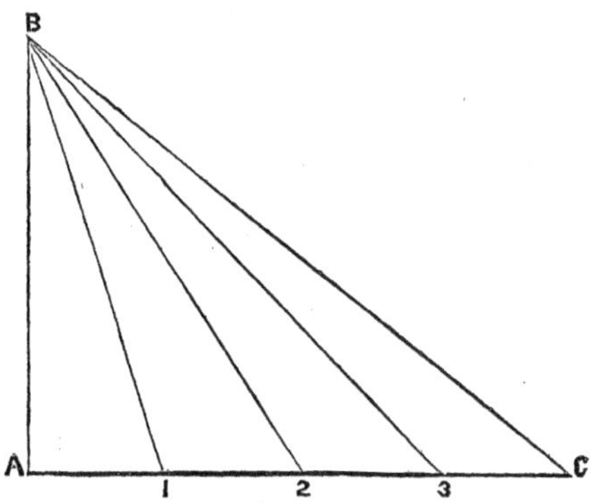

helices having the corresponding circumferences of A 1, A 2, A 3, and A C. Now, then, if these helices be the lengths of the rings, or elements for one entire convolution of the thread, all we have to do is to multiply it by the breadth of the element, which will give the area for one convolution; but as only a fraction of a convolution is used in practice, we multiply by this fraction, whatever it may be, and the product gives the area for the part used. This mode of calculation is, of course, only an approximation; but whenever the blade is divided into a considerable number of elements, say 6 inches in breadth each, the result obtains sufficiently near the truth for all practical purposes.

The following is a calculation on the screw of the U. S. Steam Frigate "Wabash," and which agrees, within a very small fraction, of the area as projected upon a plane:

Diameter of screw, 17 feet 4 inches; diameter of hub, 2 feet 4 inches.

Pitch.	Radii of Element.	Circumferences corresponding to Radii of Elements.	Lengths of Elements for one Convolution of the Thread.	Fraction of the Pitch used.	Lengths of the Elements actually used.	Breadths of Elements.	Areas of Elements.
A	B	C	D	E	F	G	H
		2.B × 3.1416	$\sqrt{A^2+C^2}$		D × E		F × G
ft.	ft.	ft.	ft.		ft.	ft.	sqr. feet.
23	1.5	9.42	24.89	2/7	7.11	.5	3.555
"	2.	12.56	26.20	"	7.48	"	3.74
"	2.5	15.70	27.85	"	7.96	"	3.98
"	3.	18.84	29.73	"	8.49	"	4.245
"	3.5	21.99	31.82	"	9.09	"	4.545
"	4.	25.13	34.07	"	9.73	"	4.865
"	4.5	28.27	36.44	"	10.41	"	5.205
"	5.	31.41	38.93	"	11.12	"	5.56
"	5.5	34.55	41.50	"	11.86	"	5.93
"	6.	37.69	44.15	14/51	12.12	"	6.06
"	6.5	40.84	46.87	"	12.86	"	6.43
"	7.	43.98	49.63	3/11	13.54	"	6.77
"	7.5	47.12	52.43	4/15	13.78	"	6.89
"	8.	50.27	55.27	1/4	13.82	"	6.91
"	8.5	53.40	58.14	1/5	11.63	"	5.815

* Helicoidal area of one side of both blades = 80.5 square feet.

Practical Remarks on the Screw Propeller.

In the application of power to the propulsion of the hulls of vessels through water, a portion of the effect is lost by the instrument through which it is transmitted. In the common radial wheel this loss of effect is compounded of two losses, slip, plus oblique action; in the feathering wheel, slip, plus drag, and in the screw propeller, slip, plus friction of the propeller blades on the water. That instrument, therefore, which, possessing no more practical disadvantages than other

* For the calculations of the friction of a screw surface on the water, see Isherwood's calculation on the "San Jacinto," (Journal of the Franklin Institute, Third Series, Vol. XXI., p. 349,) or on the "Arrogant," (Appleton's Mechanics' Magazine, Vol. I., p. 156,) from which the form for the above table is taken.

instruments, and which has the sum of its losses the least, must be the most economical propelling instrument. The feathering wheel, from what we have seen, would present itself very conspicuously to our eye as being the best instrument within our knowledge; but, unfortunately, the practical difficulties are such as to preclude its universal adoption. The loss from oblique action in the common radial wheel, particularly where the diameter is comparatively small and the dip of the paddles considerable, amounts to an important percentage of the total power of the engines; and since this loss in the screw propeller does not exist, but is replaced by one of much smaller magnitude, viz., friction of the blades, it follows, that were the slip of the two instruments alike, the screw propeller would be the more economical. In practice, however, with the screw propeller, when contending against head winds, or other increased resistance, the slip is increased to a very serious extent. In fact, in some cases it has occurred, when the engines were going ahead at nearly full speed, the vessel stood nearly still. On the other hand, however, when the sails are set to a fair wind, the slip of the propeller is materially reduced, while the thrust remains unaltered. The increased slip when contending with head winds is also experienced with paddle wheels, but they are not affected to the same extent as the propeller, the increasing or decreasing the resistance with the latter instrument, not making a vast difference in the revolutions of the engines (as is the case with the paddle wheel) so long as the pressure on the piston remains unaltered.

In the application of the screw propeller, it is well to sink it as low as possible in the water, in order that the hydrostatic pressure above may be sufficient to cause the water to flow in solid, even to the centre of the propeller, which, therefore, having the proper resisting medium, is less liable to excessive slip. This will also prevent the centrifugal action—the throwing of the water off radially from the centre—which exists to a small extent in some very aggravated cases.

Increasing the helicoidal surface of the screw beyond what is barely sufficient to transmit the power given to it, has no other effect than to occasion an increased loss by friction, by the increased surface interposed. The friction of solids on fluids, unlike solids on solids, depending upon the extent of rubbing surface as one of the elements. The object, therefore, to be sought after in practice, is to make the sum of the loss by slip, plus friction, as little as possible, and this sum, manifestly, must depend, to a considerable extent, on the amount of helicoidal surface; but, nevertheless, there appears to be no general rules yet devised, from theory or practice, which can be used as a reliable guide; different engineers making considerable difference in the areas of propellers applied to the propulsion of the same sized and modeled steamers.

Negative Slip.—It would certainly appear a very strange anomaly, were one on board a vessel, which he should discover from the indications of the log was moving actually faster through the water than the screw, there being no other propelling instrument; yet such has been apparently the case, and there are, perhaps, to this day, persons—though we hope they are very few—who think that a screw propeller may drive a vessel faster than it is moving itself. There have been cases, it is true, where the log has shown that the vessel was apparently moving faster than the screw, which alone was the propelling instrument, but that such a thing could be true is absolutely absurd, and hence attention was turned to discovering the anomaly. It is accounted for in two ways.

When a body having a blunt stern is drawn through water at a high velocity, the water, not being able to flow in from the sides of the body sufficiently rapid to fill the vacuity occasioned by its passage, flows in from all other directions, and a column of water, therefore, necessarily, follows in the wake of such a body. This is the case with screw propeller vessels having blunt runs, and, by consequence, the propeller, instead of acting upon water at rest, acts upon water in motion, having the same direction as the vessel. Now, then, supposing a propeller, acting upon water at rest, to have a slip of 10 per cent., if a column of water follow the ship with the velocity of 11 per cent. of the speed of the propeller, which still retains its ten per cent. slip, the log, as it takes no cognizance of the velocity of this water, would show a *negative slip* of 1 per cent., *i. e.*, it would show the vessel to be actually moving 1 per cent. faster than the propeller, when in reality the latter would be moving 10 per cent. the faster.

To produce such a result as this, of course, possesses no mechanical or other advantage; for power must have been originally taken from the engines to produce the current, which cannot be returned to its full extent. It is, therefore, a very important element in the design of a screw vessel to make the run very sharp—the lines fine—in order that the water may flow in solid at once, to fill the vacuity occasioned by the vessel's progress, or the propeller's revolutions.

The other theory in regard to negative slip is this: All known bodies yield to pressure, it being only necessary in order to cause the amount of yield to be measurable to make the pressure sufficiently great. It is hence conceived, that when a screw propeller is in motion, the pressure of the water on the blades causes them to spring, thereby increasing the pitch; consequently, in calculating its speed through the water, if we use the true pitch, instead of the pitch assumed, while it is in motion, the velocity given to it will be too small, and may be less than the velocity of the vessel.

We would, however, remark, that negative slip in a screw propeller, unassisted by sails, is more imaginary than real, and could only exist under very aggravated circumstances, for a screw propeller usually has about 20 per cent. slip, at least, and to reduce this to nothing, even under the conditions set forth above, would be rather a perversion of circumstances.

Altering the Pitch.

Propellers are sometimes constructed in such a manner that the pitch can be altered, from time to time, by altering the angle of the blades, which are made adjustable in a large spherical hub. Thus, if it be desired to increase the pitch, increase the angles by turning round the blades; or if it be desired to decrease the pitch, reverse the operation. Such an arrangement, however, in practice, must be confined within very narrow limits, for, inasmuch as the surface of a screw propeller blade, being that of a helicoid, every point in the blade must have a different angle, which increases as the hub is approached, and if the propeller be constructed so that all the angles be adapted to one particular pitch, it is not very likely that they will, after being distorted, be adapted to any other pitch; that is to say, if the propeller be a true screw, for instance, and have a certain angle at the periphery, if we move the blade so as to increase the angle at that point 10°, the angle at every other point in the blade will also be increased 10°, which should not be the case, but should be correspondingly less as the hub is approached; thus, by this arrangement, we give a greater pitch at the hub than there is at the periphery; and if the operation be reversed, and we decrease the angle at the periphery, the angle at the hub, and every other point in the blade, is decreased to precisely the same extent, thus giving less pitch at the hub than there is at the periphery, or any other point in the blade. We therefore arrive at this conclusion:

That having three conditions presented to us, viz., true screw, expanding screw, from periphery to hub, and expanding from hub to periphery—the latter two not in regular ratio—it is more than probable that one or the other of these must be found practically to be the superior, and whichever it may be, and that one adopted, the advantage to be derived from altering it, after it is once adopted, does not appear very plain, the arguments to the contrary notwithstanding.

Parallel Motion.

Parallel motion is a combination of bars and rods, having for its object the guiding of the piston-rod of a steam-engine in a constant straight line, or as near a straight line as can be practically attained. It is applicable, in different forms, to any type of engine, but its adaptation to the side-lever engine is the more general.

We have constructed figure 10 with the view of

Fig. 10.

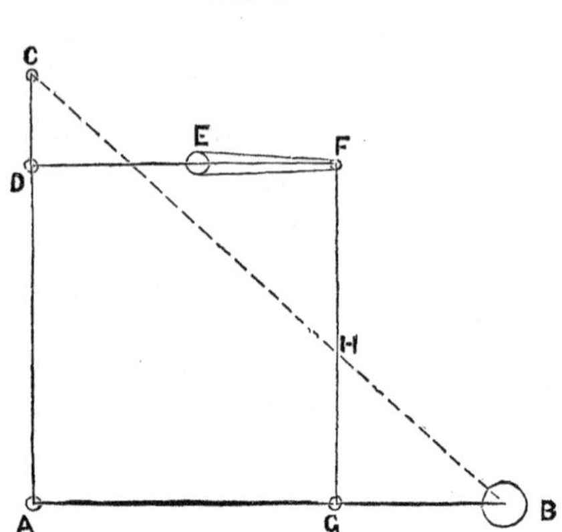

illustrating its application to this type of engine, and to clear it, if possible, of the mystery that usually hangs over it in the shape of formulas. A B is half the length of the side lever, vibrating on the centre B; A C, the side rod attached to the cross-head at C; G F, the parallel motion side rod; D F, the parallel bar, and E F, the radius bar vibrating on the centre E. The object to be attained is to make the point C travel vertically in a straight line, or as near so as possible; and from the construction of the figure, it will be seen that, when the point G moves to the right the point F moves to the left, and vice versa; hence it is manifest, that there must be some point H, in the rod F G, which will describe very nearly a straight line, and if the lengths G B and E F were equal, that point would be in the centre of F G; but, since they are of unequal lengths, H must be in such a position that $EF \times FH = BG \times GH$.

Now, then, having secured the point H, draw the line B C through H, which will determine C, the centre of the cross-head; and the triangles B H G, B C A, being similar, and joined together in such manner that, no matter how much the angles of the one may alter, the angles of the other must alter to precisely the same extent; and hence, these triangles always remaining similar, it follows that if the apex (H) of the one moves in a straight line, the apex (C) of the other must move in a straight line also.

It matters not where the points D F G may be situated, so long as D does not coincide with C, and the figure A D F G is a parallelogram; nor does it matter about the respective lengths of the sides of the parallelogram, so long as $E F \times F H = B G \times G H$.

In practice, it happens sometimes that the parallel motion gets out of adjustment, the piston rod perhaps rubbing hard on one side of the stuffing-box at the top of the stroke, and hard on the opposite side on the bottom of the stroke; or it may rub hard on the stuffing-box at one end of the stroke, and be quite free at the other. Such a result can be brought about in three ways only: either the sides of the parallelogram A D F G have got out of parallelism, the radius bar E F, of incorrect length from the wear of the brasses, &c., or the centre E has by some means been moved from its true position.

These can be all remedied by interposing liners at the proper places; of course, taking care about the centre A, in order not to endanger striking the cylinder-head, by interposing too much at that point.

Strength of Materials.

This is a subject which does not properly come within the province of the present Notes; but we have, however, thought it well to devote a short space to it at this place, confining ourselves to a few practical examples.

Beams.—The strength of beams are to each other directly as their breadths and square of their depths, and inversely as their lengths.

EXAMPLE.—The depth of the beam of an engine 75 ins. diameter of cylinder, and 7 ft. stroke, at centre is 42 ins., and using this as a standard, required the depth of one for an engine of 80 ins. diameter of cylinder, and 8 ft. stroke, the breadth, and also the maximum pressure on the steam piston to remain the same?

ANSWER.—$75^2 \times 7 : 80^2 \times 8 :: 42^2 : 2293.76$ ins. = square of the depth; the square root of which, 47.9 ins., is the depth required.

These figures, of course, do not apply to the truss, but to the solid parabolic beam.

Shafts.—The strength of shafts to resist a transverse, or torsional strain, are to each other as the squares of their diameters; for the reason that, if the diameter of a shaft be doubled, the quantity of metal is increased fourfold, which would occasion the strength to increase as the square, but at the same time there being double the leverage interposed in consequence of the double diameter, which, being multiplied by the square (or 4), will give the cube (or 8).

EXAMPLE.—The shaft of a steamer is 17 inches diameter; cylinder, 75 inches diameter; by 7 feet stroke; required the diameter of a shaft for a steamer, having an engine of 80 inches diameter of cylinder, by 8 ft. stroke, taking the shaft here given as the standard, the maximum pressure on both steam pistons to be alike.

ANSWER.—$75^2 \times 7 : 80^2 \times 8 :: 17^3 : 6388.459$, the cube of the diameter, the cube root of which, 18.55 inches, is the required diameter of the shaft.

This is about the diameter of the shafts used in practice for two engines of 80 inches diameter of cylinder, and 8 feet stroke each. The proportion in practice for a shaft for a single engine of this size, is about 15.5 ins. diameter, which is a little more than half the strength of the above shaft, owing to the weight of the wheels, &c., (which have also to be sustained by the shaft) being more than half.

Screw Propeller Shaft.—The strain on the shaft of a screw propeller is of two kinds—one in line with the axis tending to compress, the other at right angles to the axis tending to twist it. And, inasmuch as the strength of a shaft to resist compression, is much greater than that to resist torsion, we need only take the latter strain into consideration.

Hence, to ascertain the diameter of a screw shaft, the dimensions of the propeller and thrust being given, let A B, figure 11, be the pitch, B C, the circumference at centre of pressure, and A C, the helix for one convolution at centre of pressure. Draw B D at right angles to A C, and D E at right angles to B C; the lines D E and B E will be proportionally the compressional and torsional strains on the shaft; hence, if

B E be multiplied by the thrust in pounds, and divided by D E, the quotient will be the pressure in lbs., acting at the centre of pressure of the blade to twist the shaft. This pressure being multiplied into the leverage of the centre of pressure, and divided by the standard of the metal used, will give the cube of the shaft's diameter, the cube root of which will be the diameter.

FIG. 11.

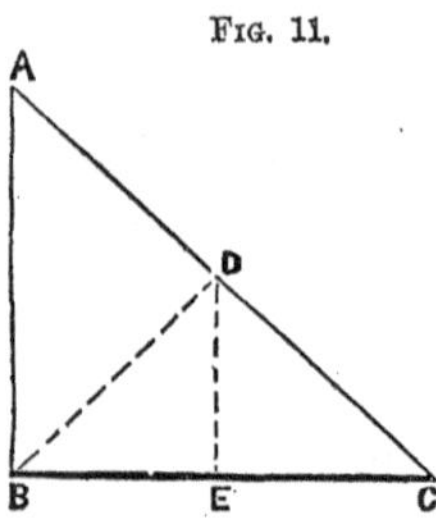

But since the triangles A C B, B D E, are similar, from the construction of the figure, the angles being respectively equal, the sides must be proportional, viz.: A C to B D, A B to B E, and B C to D E. Therefore, having the lengths of the two sides A B, B C, of the triangle A B C, we have

$$\frac{\sqrt[3]{A\,B \times t \div B\,C \times b}}{c} = \text{diameter of shaft in inches,}$$

in which A B = pitch in feet,
B C = circumference at centre of pressure in feet,
t = thrust in pounds,
b = distance from centre of shaft to centre of pressure in feet,
c = practical coefficient of the metal used for the shaft, per sq. inch of section for a leverage of one foot.

Paddle Shafts.—EXAMPLE 1.—Area of the piston 3848.4 sq. ins.; maximum pressure per sq. inch 40 lbs.; stroke 10 feet; one engine; required the diameter of the paddle shaft, the practical value of the metal being 200 lbs. per sq. inch of cross-section, with a leverage of 1 foot.

ANSWER.—$\frac{\sqrt[3]{3848.4 \times 40 \times 5}}{200} = 15\frac{5}{8}$ ins. diameter.

EXAMPLE 2.—Same as Example 1, excepting there are two engines instead of one, connected at right angles?

ANSWER.—With two engines connected at 90°, the position in which the greatest pressure on the shaft will be interposed will be when both engines are in such a position that a perpendicular, let fall from the centre of the crank upon the centre line passing through the centre of the shaft, will enclose an angle of 45°, which, with a 5 feet crank, will give a leverage of $5 \times .70711$ (nat. sin. of 45°) $= 3.535$ feet; hence, supposing the pressure at this position of the engines to be 40 lbs. per square inch, we have

$$\frac{\sqrt[3]{3848.4 \times 40 \times 3.535}}{200} \times 2 = 17.6 \text{ ins. diameter.}$$

Piston Rods.—The piston rod of a reciprocating steam-engine is subject alternately to a tensile and compressing strain; and there is nothing more absurd than the rules given in books on the steam-engine, defining its diameter to be a certain fraction of the diameter of the cylinder, independent of all other elements. For instance, suppose a rod of a certain diameter and length to be just able to sustain a certain weight placed upon the top of it, without deflexion; it is absurd to suppose that it would sustain the same weight if the rod was made double the length, retaining the same diameter; yet the rules given for the diameters of piston rods are regardless, not only of their lengths, but also of the pressure of steam. We have, therefore, thought it well to copy the following remarks and table from Johnson's translations of the book of Industrial design, by M. Armengaud, the elder, and M. M. Armengaud, the younger:

"Compression is a force which strives to crush, or render more dense, the fibres or molecules, of any substance which is submitted to its action.

"According to Rondelet's experiments, a prism of oak, of such dimensions that its length or height is not greater than seven times the least dimensions of its transverse section, will be crushed by a weight of from 385 to 462 kilogrammes to the square centimetre of transverse section, or a weight of from 5470 to 6547 per square inch of transverse section.

"In general, with oak or cast iron, flexure begins to take place in a piece submitted to a crushing force, as soon as the length or height reaches ten times the least dimension of the transverse section. Up to this point the resistance to compression is pretty regular.

"Wrought iron begins to be compressed under a weight of 4900 kilog. per square centimetre, or of nearly 70000 lbs. per square inch, and bends previously to crushing, as soon as the length or height of the piece exceeds three times the least dimension of the transverse section."

We show, in the following table, to what extent per square inch we may safely load bodies of various substances:

Table of the Weights which Solids—such as Columns, Pilasters, Supports—will Maintain without being Crushed.

WOODS AND METALS.

Description of Material.	Proportion of Length to Least Dimensions.				
	Up to 12.	Above 12.	Above 24.	Above 48.	Above 60.
	lbs.	lbs.	lbs.	lbs.	lbs.
Sound Oak	426.750	355.625	213.375	71.125	35.562
Inferior Oak	270.275	119.490	71.125		
Pitch Pine	533.437	440.975	266.007	106.687	
Common Pine	137.982	116.645	69.709		
Wrought Iron	14225.000	11877.875	7112.500	2375.575	1994.900
Cast Iron	28450.000	23755.750	14225.000	4741.666	2375.575
Rolled Copper	11707.175				

Example.—What is the least diameter of a piston rod for a cylinder having a cross-section of 3848.4 square inches, to sustain with safety a pressure per square inch of piston of 40 lbs., the proportion of length to be about 24 to 1?

Answer.—Taking one half the number in the above table for the practical value, we have

$$\frac{3848.4 \times 40}{7112.5 \div 2} = 43.28604 \text{ sq. ins. cross section of the rod,}$$

$$\text{and } \sqrt{\frac{43.28604}{.7854}} = 7.4 \text{ ins. diameter of the rod.}$$

Surface Condensers.

A surface condenser is an instrument for condensing steam by contact with cold metallic surfaces, instead of bringing it directly into contact with a shower of cold water. The object of using such a condenser in lieu of the common jet, is to furnish boilers of marine steamers with distilled instead of sea water, consequently to provide against the loss of fuel otherwise occasioned by blowing off a portion of the water, to keep the concentration at a desired point, as shown at pages 66 and 67. Also to prevent the loss due to the little conducting power of the envelope of scale which attaches to all heating surfaces of boilers using sea water.

By the use of such an instrument there is also gained the saving in labor of scaling and cleaning the boilers, which belongs to all sea steamers using the common jet, and this is of no small importance to those having the care of steam machinery.

Again, by its use the expense of repairs to the boilers is considerably reduced, their durability greatly increased, the pressure of steam which can be judiciously carried is unlimited, and the expansion of the steam can be carried to a greater extent.

With these many marked advantages, it seems extraordinary that the introduction of surface condensers should have met with so little encouragement; the slow progress made has not been owing to any want of engineering ability, but solely for the want of patronage; for engineers of talent both here and in Europe have devoted their time to the subject for many years, and have produced many forms, some of which have been so successful as to render, in our opinion, the use of jet condensers absurd. Of the number invented and introduced into practice, the one known as Pirsson's has thus far met with the most favor. It is termed a double vacuum condenser, *i. e.*, it has a vacuum within and without the condensing tubes. The injection water is received upon a scattering plate, and showered down on the tubes, which condenses the steam within them; this injection water with the air and uncondensed vapor is extracted by an air-pump, in the same manner as when the jet condenser is used, and the water of condensation is drawn away by a separate pump, called the fresh water pump, and discharged into a reservoir, whence it is delivered by the feed-pumps into the boilers.

Another variety of condenser, known as Sewell's, has recently attracted considerable notice. It has been highly reported upon by a Board of naval engineers appointed by the Hon. Secretary of the Navy, and has been introduced into some of the most successful steamers. It is of the close surface type, that is, it has a vacuum upon only one side of the condensing tubes, the condensation being effected by currents of cold water driven through the tubes by a pump. The joints of the tubes are made with india-rubber sleeves, so that they give perfect tightness, and allow each tube to expand or contract by itself, independent of the others, and each or all of them can be taken out for cleaning or repairs. The vacuum produced by this condenser is unequalled, and as there is but one air-pump, it is obtained with less power than by the other method.

Close tube surface condensers had been made with the tubes secured at both ends without any provision for expansion or contraction, except the buckling of the tubes when hot, and stretching when cold. They have also been made with one end of the tubes only secured, the other ends being fitted to an expansion plate. The advantages of Mr. Sewell's over the latter named plans are manifest.

The following figure* will give the student a clearer idea of the construction and operation of Pirsson's condenser. A A, is the condenser, in which there is a

* Taken from "*Steam for the Million*" by Commander WARD, U. S. N.

series of small tubes: *p*, the air-pump; *f*, fresh water-pump; *b*, the exhaust pipe; *l*, the injection pipe. The

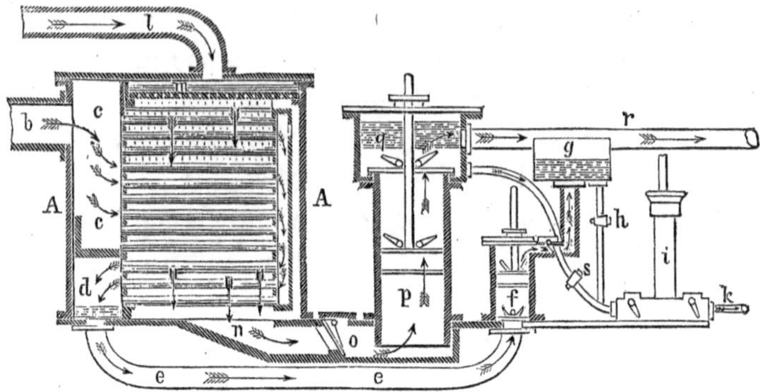

operation is as follows:—The engine being put in motion, the exhaust steam flows through the exhaust pipe *b*, into the chambers *c c*, thence in direction of the arrows through the tubes to the lower chamber *d*, injection water being admitted at the same time from the sea through the injection pipe *l*, is showered by the scattering plate *m* over the tubes, and by its gravity takes the direction of the arrows to the channel way *n*, from which it is removed by the air-pump *p*, and delivered into the hot well *q* to the delivery pipe *r* and overboard.

The water resulting from condensation is drawn by the fresh water pump *f* from the chamber *d*, through the pipe *e e*, and delivered into the fresh water reservoir *g;* from this reservoir it passes to the feed pump *i*, through the pipe *h*, and is delivered into the boilers through the pipe *k*. The pipe *s* is for the purpose of supplying salt water when deficiencies occur.

In this condenser, as drawn, all the tubes are firmly secured to both tube heads, but one end of the tube box is free, so that all the tubes can expand and contract together; those recently constructed have each tube secured to the tube head at one head only, the other ends being fitted so that they just pass through the holes, thus allowing each tube to expand or contract regardless of the others. There is also a communication from the exterior to the interior side of the tubes, so that the vacuum within created by the fresh water pump is equal to that without, created by the large air-pump. In close tube surface condensers, the position of the steam and water, as shown in the figure, is reversed. The exhaust steam is received on the exterior of the tubes as at *l*, and is condensed by water entering at *c′ c*, and driven through thetubes by a circulating pump, attached at *b*; it is then discharged through a pipe from *d*. The pump *p* is converted into a fresh water air-pump, receiving the fresh water through the channel-way *n* and foot-valve *o*, and discharging it into the reservoir *q*, whence it is received by the feed-pump and pumped into the boilers.

Cylindrical Boilers.

The force tending to rupture a cylinder along the curved sides depends upon the diameter of the cylinder and pressure of steam, and we may regard, hence, the total pressure sustained by the sides to be equal to the diameter × pressure per unit of surface × length of boiler, neglecting any support derivable from the heads, which, in practice, depends on the length. The shorter the tube, the greater its powers of resistance. This is in

Fig. 12.

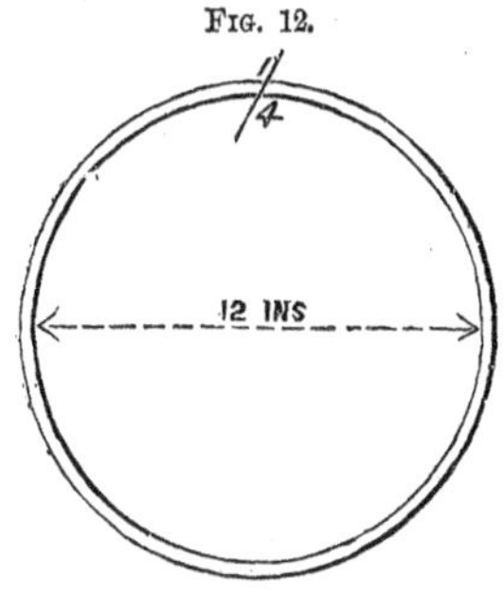

consequence of the ends being rigid and unyielding.— See latest experiments on this subject by William Fairbairn, Esq., C. E., F. R. S.

The force tending to rupture a boiler is termed, by Professor Johnson, the *divellant* force, and the tenacity or strength of the metal which resists the *divellant* force is termed the *quiescent* force. When rupture is about to take place, these two forces must be exactly equal.

Example.—What pressure will a cylinder boiler, 12 ins. diameter, and ¼ in thickness of metal, sustain per square inch, the iron to be of the best English iron?

The experiments of the Franklin Institute give for the strength of single riveted seams, 56 per cent. of the sheet, and assuming the tensile strength of the best English iron to be 60,000 lbs. per square inch of section, we have

$$\frac{60{,}000 \times .56}{12\,(\text{diameter}) \times 4\,(\text{length of band to make 1 sq. in. area of cross section})} = 700 \text{ lbs. per sq. in.}$$

But as the opposite side of the boiler will support an equal amount, the true pressure will be double this, or 1400 lbs. per square inch, one-fourth of which only (350 lbs.) would be safe to subject it to in practice.

From this we see that the bursting pressure of a boiler of the dimensions above given, in a transverse direction, is 1400 lbs. per square inch. We will now see what force this 1400 lbs. exerts to tear the boiler asunder in a longitudinal direction. To do this, we have only to multiply the area of the head by the pressure per square inch, and divide by ¼ the circumference, (since the iron is ¼ inch thick,) which will give the strain upon each square inch of sectional area.

Thus $\frac{113.09 \times 1400}{37.69 \div 4} = 16800$ lbs. per square inch of sectional area, in a longitudinal direction, and $\frac{1400 \times 12 \times 4}{2} = 33600$ lbs. per square inch of sectional area in a transverse direction.

The 4 in the latter case is the length of the band to give one inch square of sectional area, and we divide by 2 because there are two sides of the boiler to support the pressure.

From these figures, it is observed that the strain upon a cylindrical boiler, or other cylindrical vessels, subject to internal pressure, transversely, is exactly double what it is longitudinally. In cast iron, or other cast metal cylindrical vessels, this is made amends for, in a certain degree, by casting ribs, or bands, around the external surface; but with boilers there appears to have been no attempt to increase the strength by riveting bands at intervals on the outer surface, though we see no good reason why such a thing could not be done very advantageously.

We remark, from what has appeared, that the strain upon cylindrical boilers increases transversely directly as the diameters, and longitudinally as the squares of the diameters—because the areas of the heads increase in that ratio—but the circumferences increase also as the diameters; and hence, though we obtain four times the pressure longitudinally by doubling the diameter, we have double the metal in the circumference of the boiler to sustain it, and, therefore, the strain upon a unit of metal, in this direction, increases also as the diameter. Hence, no matter what may be the diameter of a boiler, the transverse pressure tending to tear it asunder, will always be double the longitudinal pressure.

Boiler Explosions.

There is only one grand direct cause of boiler explosions, and that is the incapacity of the metal, at the time, to sustain the pressure to which it is subjected. This can be brought about in several ways; defective material of which the boiler is constructed, defective construction, all parts of the boiler being incapable of sustaining the same pressure, gradual accumulated pressure without the means of escape, sudden accumulated pressure occasioned by pumping water on red-hot sheets, collapse occasioned by a vacuum in the boiler, the reverse valve being inoperative; collapse of flue occasioned by internal pressure in the boiler and a partial vacuum in the flue; overheating the plates, brought about by the accumulation of large quantities of scale upon them, thereby reducing their tenacity.

Boilers having been previously tested by hydrostatic pressure considerably beyond the limit to which it is intended ever to allow the steam to reach, and each and every boiler being fitted with steam and water-gauges, proper sized safety-valves and such like instruments, there is never any good excuse, under any circumstances, for the cause of boiler explosions. Incompetency or recklessness must be somewhere manifest, for the engineer, knowing the pressure which his boiler will with safety bear, should under no circumstances allow it to exceed that pressure. We would, however, observe here, that we have noticed in many cases, both ashore and afloat where there are a number of boilers connected together, instead of having a steam gauge attached to each one separately, there was but one gauge to the whole number; and hence, if one or more boilers be shut off from the others, there would be no means of ascertaining the pressure within them; and it is a very common thing with land boilers and boilers of small river boats to have no steam-gauge whatever. In such cases as these the owners take upon themselves the responsibility, which would otherwise be attached to the engineer, of any disastrous result.

The legislation in regard to the inspection of steam-boilers is hardly adequate to the cause; for though the testing the strengths of boilers, from time to time, is very good as far as it goes, it falls short of what the seriousness of the case demands. The same amount of strict, unbiased inspection on the parties who have charge of the very powerful, yet governable element of steam, would be followed by far more beneficial results. Place only those in charge of the steam-engine, boilers, and dependencies, who are competent to the task; prevent owners from employing any one simply because his services can be secured for a small compensation, and then you touch the subject in a vital point. It is too prevalent an opinion, that any one who can stop and start an engine, have the fires started and hauled, is an engineer, regardless of his knowledge of the element of which he has charge.

It is true, however, that the system of rivalry and competition, carried on by steamboat owners and others using steam power, is such as to prevent any one independently from paying a very high rate of compensation; but if all were compelled to employ equally competent services, no difficulty could be experienced on this head.

Horse Power.

The standard for a horse power in England and the United States is pretty generally established at 33000 lbs. raised one foot high in a minute; but in France a horse power is estimated at 75 kilogrammètres, which is 75 kilogrammètres raised one mètre high per second, equal to 32554.7 lbs. avoirdupois, raised one foot high per minute. To ascertain the horse power of a steam engine, *multiply the mean unbalanced pressure per square inch on the piston, by the area of the piston in square inches, by the length of the stroke in feet, and by the number of strokes in a minute; and divide by* 33,000, *the quotient will be the horse power.*

From this figure, in order to ascertain the actual power utilized in propelling the vessel, a deduction has to be made for working the air and feed pumps with their load, friction of working journals, friction of load on working journals, amounting in all to about 20 per cent. of the total power, leaving 80 per cent. to be applied to the propelling instrument, which 80 per cent. has to be reduced by the amount of loss which obtains in the propelling instrument.

Example.—Required the horse power of a condensing steam-engine, having a cylinder 70 inches diameter, by 10 feet stroke, making 15 revolutions per minute; mean pressure of steam throughout the stroke 23 lbs.; back pressure 3 lbs.; and also the actual power utilized in propelling the hull of the vessel, the sum of losses in the propelling instrument being 40 per cent. of the power applied to it?

Answer 1st.—

$$\frac{70^2 \times .7854 \times 23 - 3 \times 10 \times 15 \times 2}{33000} = 699.7 \text{ horse power.}$$

Answer 2d.—Considering 20 per cent. of the total power to be expended in working pumps, in friction, &c., we have 80 per cent. applied to the propelling instrument, and 40 per cent. of 80 per cent. = 32 per cent. of the total power expended in transmission through the propelling instrument; wherefore, 80 − 32 = 48 per cent. of the total power applied to propelling the hull of the vessel = 335.856 horses.

Nominal Horse Power, is a term which expresses neither the actual power, the size of the engine, nor any thing else which is useful; and though it has become almost obsolete among well-informed engineers in this country, our trans-atlantic friends seem yet to cling to it with some tenacity.

The usual rule for determining it is this: *Multiply the square of the diameter of the cylinder in inches, by the cube root of the length of the stroke in feet, and divide by 47; the quotient is the horse power.*

Now, the chief object for establishing a rule for nominal horse power was to create a *commercial* unit, by which the power of one engine could be compared with that of another engine; and this rule might meet the wants of the case, did the lengths and breadths of all cylinders bear the same ratio, and did the pressure of steam remain an invariable quantity: but as these elements are constantly varying, it is of no use whatever; and further, if they did not vary, the simple square of the diameter would express an unit equally incorrect. In order to show further the utter uselessness of the term horse power, as expressed above, we will take two engines, each having 70 inches diameter of cylinder, one 10 feet stroke, the other 5 feet stroke, and ascertain the nominal horse power of each.

$$\frac{70^2 \times \sqrt[3]{10}}{47} = 224.7 \text{ horses.}$$

$$\frac{70^2 \times \sqrt[3]{5}}{47} = 178.2 \text{ horses.}$$

Now, then, if the pressure of steam was the same in these two cylinders, and the pistons moved with the same velocity, it is manifest that the powers must be the same; yet, according to the rule for nominal horse power, they are made widely different; and if so much difference is made while the pressure of steam is supposed to remain constant, what must we expect when that element also varies?

Vibration of Beams.

Given, the length, O *c*, from centre of beam, to *a b*, line passing through centre of cylinder = 10 feet; and

Fig. 13.

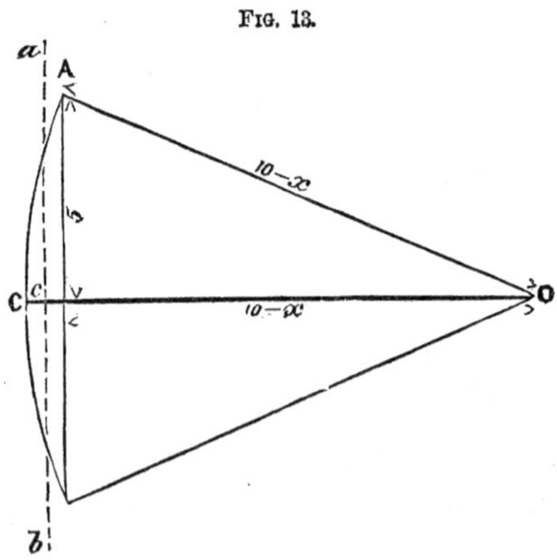

length of stroke = 10 feet; required, the length O A, or O C, of half the beam?

The line $a\ b$ bisects the versed sine of the arc, and, supposing one half ($c\ C$) of the versed sine to be $= x$, we have $(10 + x)^2 = (10 - x)^2 + 5^2$

$$100 + 20x + x^2 = 100 - 20x + x^2 + 25$$
$$20x + 20x = 25$$
$$x = .625,$$

Hence, half the length of the beam $= (10 + .625) = 10.625$ feet.

Marine Economy.

A body moving through water with a certain velocity displaces a certain quantity of water in a given time, with a certain velocity; if the velocity be doubled, the quantity of water displaced will also be doubled, because the body moves double the distance, and each particle of water will, therefore, be displaced with double the velocity; hence, the resistance to the body will be as 2×2, or as the square of the velocity. Thus it appears that, if a ship consumes 500 tons of coal to perform a certain distance, at the rate of 5 miles the hour, to perform the same distance at the rate of 10 miles the hour, would require $5^2 : 10^2 :: 500 : 2000$ tons, or 4 times 500; but the quantity of coal required for any one day, at the rate of 10 miles, will not be 4 times the quantity required at that rate for 5 miles, but will be 8 times; for, supposing the speed be increased to 10 miles the hour, the same distance will be performed in 5 days; hence, we have, in the first case, 500 tons consumed in 10 days $=$ 50 tons per day, and in the latter case, 2000 tons in 5 days $=$ 400 tons per day, or 8 times 50 tons. Now, then, taking the coal as the exponent of the power, we see that the power has to increase as $2 \times 2 \times 2$, or as the cube of the velocity. Hence the importance, wherever speed is not an object, of running the engines as slow as possible, in order to economize the fuel.

But whenever there is an adverse current to contend with, the most economical speed is *half as fast again as the current.* That is to say, if the velocity of the current be 4 miles the hour, the velocity of the vessel should be 6 miles. We will endeavor to demonstrate this without the use of mathematical formula.

Let 1 represent the power required for a speed of one mile per hour, then, inasmuch as the power increases as the cube of the velocity, the power required for the speed of 6 miles $= 6^3 = 216$, and the ground moved over $= 4 - 2 = 2$.

Suppose, now, the velocity of the ship be reduced to 5 miles per hour, the power will be $= 5^3 = 125$, and the ground moved over $= 5 - 4 = 1$.

Suppose, again, the speed to be increased to 7 miles per hour, the power will be $= 7^3 = 343$, and the ground moved over $= 7 - 4 = 3$.

Summing up these figures, we have for a speed of 7 miles per hour a power expended of 343, to make good a distance of 3 miles $= 114\frac{1}{3}$ per mile; for a speed of 5 miles, a power of 125 to make good 1 mile $= 125$ per mile; and for a speed of 6 miles, 216, to make good 2 miles $= 108$ per mile. Consequently, the least power is required at the speed of 6 miles, which is half as fast again as the current.

Had the calculation been made for any fraction of a mile, either above or below 6, the same result would have been obtained.

These calculations apply alike to head winds, &c., at sea, as well as to a tide-way in a river; whence it follows that a vessel can be run even too slow for economy, but nevertheless, when having a heavy head sea to contend with, there are other elements besides economy of fuel to be taken into consideration; the strain upon the vessel and machinery, the plunging and staving in of the light work about the bows and other places, shipping of seas, &c., are matters which also require the judgment of the commanding officer.

Limit to Expansion.

Theoretically, supposing a perfect vacuum to obtain in the cylinder, there is no limit to expansion; but, practically, there is. The unbalanced pressure at the end of the stroke should never be less than sufficient to overcome the friction of the engine, and ought always to be a little more.

EXAMPLE.—Length of stroke = 8 ft.; initial pressure of steam 30 lbs. per square inch, inclusive of the atmosphere; back pressure 4 lbs. per sq. inch; friction of engines, &c., = 2 lbs. per square inch; required the point where the steam should be cut off to yield all its useful effect?

$$x = \text{the point,}$$
$$4 + 2 = 6 = \text{the pressure at the end,}$$

$$x \times 30 = 6 \times 8$$
$$30x = 48$$
$$x = 1.6 \text{ ft. from commencement.}$$

The Proper Lift for a Valve

Is equal to the area of the valve divided by the circumference.

And the metal of which they are made is either brass or cast iron; the latter has the advantage of expanding nearer equal with the steam chest.

Temperature of Condenser.

Example.—Water in the boilers, carried at a density of $1\frac{3}{4}$ per saline hydrometer; temperature of the condenser, and water entering the boilers, 105° Fahr.; vacuum in condenser, 27.82 inches. Compare the economic performance of the engine, under these circumstances, with the same engine, carrying the water in the boilers at the same density, but the water in the condenser at 120° Fahr.; the mean pressure of steam in both cases on the piston being 20 pounds per square inch?

Solution.—Neglecting the difference of power in the two cases required to work the air-pump, taking the boiler pressure at 20 lbs., and 2 inches of mercury to be equal to 1 lb. pressure, we proceed thus:

$1184-105\times.75+228.5-105 : 228.5-105 :: 100 :$ 13.23 per cent. loss by blowing off, in the first case.

$1184-120\times.75+228.5-120 : 228.5-120 :: 100 :$ 11.96 per cent. loss by blowing off, in the second case.

$20\times2 : 2.18$ (back pressure) $:: 100 : 5.45$ per cent. of the effect of the engine lost by back pressure, in the first case.

$20\times2 : 3.33$ (back pressure) $:: 100 : 8.325$ per cent. of the effect of the engine lost by back pressure, in the second case.

Now, then, letting the fuel represent the power, we observe, in the first case, that only (100−13.23=) 86.77 per cent. reaches the engine, of which 5.45 per cent. is lost in back pressure, and 5.45 per cent. of 86.77 per cent.=4.73 per cent. of the total effect lost by back pressure, leaving (86.77−4.73=) 82.04 per cent. to be applied to operating the engine.

In the second case, (100−11.96=) 88.04 per cent. of the power reaches the engine, of which 8.325 per cent. is lost in back pressure, and 8.325 per cent. of 88.04 per cent.=7.33 per cent. of the total effect lost by back pressure; leaving (88.04−7.33=) 80.71 per cent. to be applied to operating the engine.

Therefore, under the conditions of the example, the engine, in the first case, performs the same amount of work with (82.04−80.71=) 1.33 per cent. less fuel.

This calculation can be made accurate by taking diagrams from the cylinder and air-pump, under the conditions of the example, and estimating the power in each case; then, the power to work the air-pump is considered.

CHAPTER VI.

WESTERN RIVER BOAT ENGINE.

FIG. 1 is a side elevation, and Fig. 2 an end view of the celebrated high-pressure engine, so extensively employed on all the Western rivers, also in many iron-rolling mills, and other manufacturing establishments of the West.

The first steamboat that ever navigated the great rivers Ohio and Mississippi was called the New Orleans, built at Pittsburgh, Pa., by Mr. Rosevelt, for Messrs. Fulton & Livingston, of New York; launched March, 1811, and made a passage to New Orleans the latter part of the same year. We have no reliable information in regard to the kind of machinery used on board that boat. The type at present employed, however, came into use early in the history of Western river steam navigation. It subsequently underwent several modifications, but for a quarter of a century it has been made essentially as represented by the following cuts. In fact, so alike have they been made, for many years, that those engaged in constructing them make no drawings, but continue from year to year to cast from the same patterns, and make and erect without variation.

In all the side-wheel boats, the engines are disconnected, separate, and distinct, so that by revolving the wheels in opposite directions, the boats can be turned on a pivot.

Fig. 1.

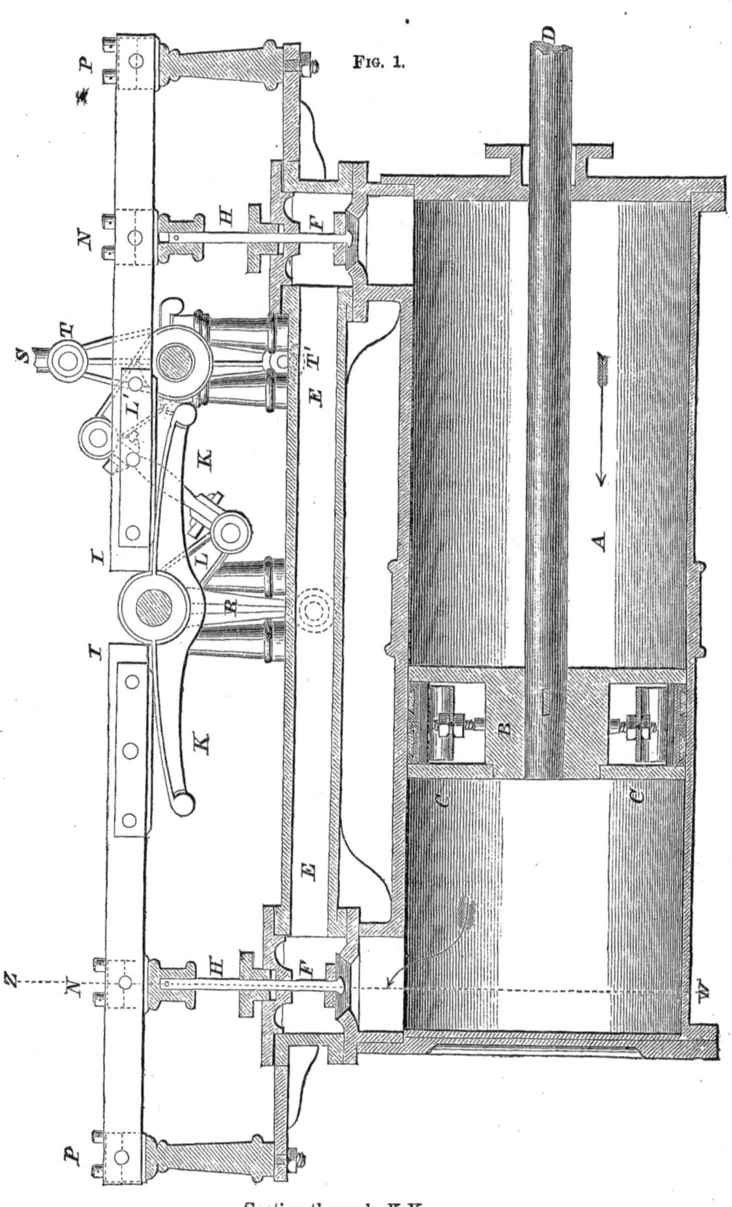

Section through $X\ Y$.

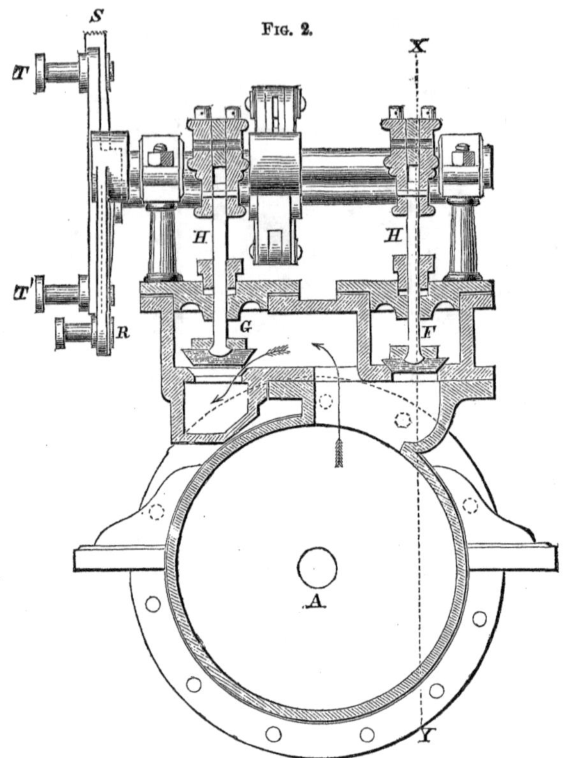

Section through W Z.

EXPLANATIONS OF DIAGRAMS.

The piston is represented to be proceeding in the direction indicated by the arrows, both steam valves closed and one exhaust valve open; the steam is therefore cut off, and acting expansively.

Like letters refer to like parts in both views:

A, steam cylinder; *B*, steam piston, with metallic packing set out with springs, as recently introduced; *c c*, piston follower; *D*, piston rod; *E E*, steam side pipe, to which the steam pipe from boilers is attached; *F F*, steam valves; *G*, exhaust valve; *H H*, valve stems; *I I*, valve levers; *K K*, steam valve lifters; *L L'*, arms connecting steam rock shaft with exhaust lifters, on opposite side of engine, shown in Fig. 2; *N N*, points of application of weight or pressure to levers; *P P*, fulcrums of levers; *R*, steam rock shaft arm, to which cut off cam rod is connected; *T T*, pins of rock shaft arms, to which full stroke cam rod is hooked; *S*, starting bar.

The lifters for working the exhaust valves, with arm *L* cast on, are loose, and vibrate on the steam rock shaft. They are on the opposite side of engine, as shown in Fig. 2. The arms *L L* are of the same length, being used only to connect the exhaust rock shaft to the exhaust lifters.

The valves are worked by cams; the exhaust cam being made for full stroke, and the steam cam to cut off the steam at defined points of the stroke.

There is a short shifting link to connect the exhaust rock shaft arm at *T* with the steam rock shaft arm *R*, so that when it is desired to work the steam full stroke, instead of expansively, it is only necessary to unhook the steam cam rod from the pin on the arm *R*, and hook on the short link; thus connecting all the valve levers, and working all the valves by the full stroke cam. To back the engine, the engineer has only to shift the full stroke cam rod from the lower pin *T* to the upper pin *T*. The laborious working of the engines by hand is therefore entirely avoided; and as there is much backing to be done at the different landings, this is of importance.

The pitman (connecting rod) is of wood strapped with iron, and from 3 to 4 times the length of stroke.

The valves are of cast iron, sometimes double and

balanced, but most frequently single and unbalanced, as represented in the diagrams. In the latter case, considerable power is expended in working them against so high a pressure, unless the valve gear be constructed to reduce the power to a minimum at the point of its application. This, however, is not often done, though it can be so arranged. And, as a matter of exercise for the student's mind, we will give some explanation of it, and of the operation of the valves.

The gear may be considered a compound lever, or two connected levers, transmitting the power from the cam rod pins through each other to the point of resistance, weight, or pressure. The explanation, then, is as follows: Call lever *I* No. 1, and lifters *K K* lever No. 2; say the length of the long arms of the levers are 64 and 9 inches, and the short arms 16 and 3 inches. That is, the distance from fulcrum *P* of lever No. 1 to the end or point where the lifter first touches it and begins to raise the valve is 64 inches, and the distance from centre of steam rock shaft, which is the fulcrum of lever No. 2, to cam rod pin centre is 9 inches; also that from *P* to *N*, of lever No. 1, is 16 inches, and from centre of steam rock shaft of lever No. 2 to point on lifter where it commences to raise the valve lever is 3 inches. Now, suppose there to be a pressure of 140 lbs. per square inch in the pipe *E E*, consequently on the valves, and that the diameter of the valve is 6 inches, there is hence a pressure on each valve of $6 \times 6 \times .7854 = 28.274 \times 140 = 3958.36$ lbs., which, added to weight of valve and lever, weighed at *N*, say 40 lbs. more, $= 3998.36$. This will be reduced at *I*, end of lever No. 1, to $3998.36 \div 4 = 999.59$ lbs., and at the pin of steam rock shaft arm to $999.59 \div 3$

= 333.196 lbs. required to operate the valves, friction not included. Or the work may be shortened by multiplying all the long arms together, all the short arms together, dividing one by the other, and the weight by this result, thus:

$$\frac{64 \times 9}{16 \times 3} = 12, \text{ and } 3958.36 + 40. \div 12 = 333.196. \text{ lbs.}$$

To work the engine by hand, this weight can be reduced by the length of starting bar to say 54.97 lbs. And it can be further reduced by closing the exhaust valves sooner, so as to partially balance the steam valves: for instance, suppose they be closed when the piston is in the position to leave 6 inches between it and the cylinder head; that there is a clearance of ¾ of an inch, and that the back pressure against the piston is 5 lbs. per square inch when the valve is closed; now, according to the law of expansion and compression of gases, when the piston has travelled 3 inches further, or half the distance from point of closing the valve to cylinder head, there will be a pressure against it of 10 lbs. per square inch, and when it has travelled 1½ inches further, there will be a pressure of 20 lbs. per square inch, and when ¾ of an inch further, or at end of stroke, there would be a pressure of 40 lbs. per square inch for the piston to cushion against, but the valve is not relieved to this extent if it has any lead; say it is opened 1½ inches before the piston arrives at the end of its stroke, we then have only 20 lbs. per square inch pressure under the valve to be deducted from the 140 lbs. per square inch above it, or a total steam pressure on the valve of 3390.08, instead of 3958.36.

We have been considering the power necessary to

work each steam valve separately: we will now consider that requisite to operate each exhaust valve. The proportion of levers and gear remaining the same, the steam cut off at half stroke, and the diameter of valve 7½ inches—this being the ratio considered necessary for the free and quick escape of the steam—there will consequently be by the reduction of pressure common to expansion, at the time the valve is to be opened, 70 lbs. per square inch on it, or in round numbers 1979.18 lbs. less than on the steam valve. To cut off shorter, this pressure is still further reduced; to follow longer, it is increased—the reduction of pressure and temperature by leakage and condensation in the cylinder not being considered.

This type of engine is peculiarly adapted to the boats on which it has been exclusively employed for more than quarter of a century; and when well constructed, correctly proportioned, and properly managed, it performs the work satisfactorily, can be handled with facility, is durable, and could be made economical. But all these elements are seldom found in any of them. The valve gear is generally arranged so that the valves cannot be worked by hand against the full pressure on them; the steam valves are set with little or no lead, and the exhaust valves do not close until the piston arrives at the end of the stroke, so that there is only a very small cushion against the pistons at the end of the cylinders, and the cranks pass the centres against an unbalanced pressure of from 50 to 60 lbs. pressure per square inch. The cams for working the steam valves are usually made to cut off the steam at half, five-eighths, or three-quarters from commencement of stroke; and the cylinders, steam pipes and drums, upper portions of boilers, etc., are always uncovered (not jacketed) by non-conducting substances, and exposed to the cold winds sweeping between the decks. It will therefore be seen that a large margin is left for improvement in proportions, and more so for saving fuel. The temperature of the steam in boilers, pipes, etc., may be averaged at 360° Fahrenheit, and the temperature of the atmosphere during the year at 60°; hence there is a difference between the temperature of steam within the vessels and the atmosphere outside of 300°; and considering the unusual large radiating surface exposed to the winds and cold air, the loss from the condensation of steam in cylinders, pipes, etc., cannot be less than 15 per cent. of the fuel consumed. And the saving which could be realized in fuel from a high degree of expansion, where the pressure is about 140 lbs. per square inch, may be estimated by the student from calculations under that head commencing at page 12 of this work.

The lifter represented in the diagrams was introduced not very long ago, and patented by Mr. A. Hartuper, of Pittsburg, Pa. It is a great improvement over the lifter formerly employed and still used to a great extent.

The improvement consists in commencing to lift the valve lever close up to the rock shaft, and starting with an easy curve from the horizontal centre line of the rock shaft downward. Its advantage is to be found in a smooth-working valve gear, and reduced power to work the valves consequent on the reduced distance between the fulcrum of lever No. 2 and application of power on lever No. 1; that is, between the centre of rock shaft and point where the lifter first touches the lever and begins to raise the valve. The valve once started from its seat, steam rushes under it, and assists its ascent; hence, as the bearing point of the lifter approaches its end, the valve becomes balanced from the steam under it. Furthermore, the shape of the lifter is such as to raise the valve quickly.

The ordinary lifter begins to raise the valve lever from and near its end, and several inches from the end of lever *I*, instead of near the fulcrum of lever No. 2, and at the end of lever No. 1. It will therefore be readily seen that, if this lifter should be substituted for the one represented in the drawing, the power to work the valves would be greatly increased. As an illustration, suppose the arms *K K* to commence lifting 12 inches from centre of rock shaft, and 9 inches from end of lever No. 1, then there would be a power of 1160.9 lbs. required at *I*, and 1546.78 lbs. at the end of arm *R*.

In addition to this increased power to work the valves, the sudden striking of the valve levers by the old-fashioned lifters is objectionable, and involves the necessity of putting a leather shoe on the lever to ease the disagreeable noise.

In explanation of the reduction of power consequent upon the combination of levers, the student must bear in mind that what is gained in power is lost in time, the lifters being but a short space of time raising the valves, during the stroke of the engine.

The kind of lifter represented in the diagram, Fig. 1, but with much greater curve downward, is sometimes used on land engines, and operated by the eccentric to cut off the steam at defined points of the stroke, in the same manner as the Stevens cut off; namely, by lost motion, or, in other words, by causing the toes to travel a considerable distance before commencing to raise the valves.

In order to get a clear understanding of this, the student can refer to the diagram of the Stevens cut off, page 22, and consider the steam-lifting toes *B B*, in that diagram, in the same light as the valve levers *I I* of Fig. 1 in the above. The principle on which the two kinds of cut offs are operated is precisely alike, the only difference being that one is made adjustable and the other is not.

Some of the boats plying on the upper rivers have stern wheels, *i. e.*, one paddle wheel applied at and projecting over the stern the whole width of the boat. Many of such of 300 tons burthen draw only from 16 to 20 inches of water: lightness of machinery, compatible with strength, is therefore of the first importance in such vessels. The boilers of all of them are, as a rule, placed near and fronting the bow; and as the stern wheel variety involves the necessity of placing the engines near the stern, the steam pipes are as a consequence exceedingly long, not unfrequently from 90 to 100 feet or more, thus causing still further loss from condensation of the steam. It is proper, however, to remark, that when coal can be furnished, as it is on the upper rivers, at less than one dollar per ton, economy of fuel is a secondary consideration. It is also proper to state that the owners and operators of the boats are slow to make improvements in their steam machinery, even when the advantages of such can be practically demonstrated. As an instance of this, we may mention that spring cylinder piston packing, so long successfully applied to the pistons of all types of engines, is now, the year 1862, introduced in the river engines for the first time.

It is to be regretted that the Indicator, engine registers, and correct guages have not come into use, and records kept on board of some of the boats, so that correct data for calculating their efficiencies and comparative economy could be had. If we are not misinformed, Mr. S. H. Gilman is the only engineer who ever applied the Indicator and made experimental tests for the purpose of getting correct results from the Mississippi steamers. Among the few records kept by him, we select for the benefit of those interested that of the "Magnolia," one of the finest and largest steamers plying on the lower Mississippi when the record was taken, namely, in the year 1853, and published immediately afterward in the "Franklin Institute Journal."

The diagrams taken from the cylinders of that vessel showed the valves to have been set without lead, and that the engines passed the centres with the unbalanced pressure of 60 lbs. per square inch; the stroke being 10 feet, and the steam following the pistons 6 feet before being cut off—that is to say, both the steam and exhaust valves opened and shut precisely at the end of the stroke, and that the steam was expanded down to 60 lbs. pressure above the atmosphere when the pistons reached the ends of the cylinders.

DIMENSIONS AND PROPORTIONS OF THE MAGNOLIA.

Length from stem to stern	295 feet.
Breadth of beam	35 "
Breadth of floor	28 "
Depth of hold	9 "
Draft of water when light	4 "
Tonnage, carpenters' measurement	914 tons.
Diameter of water wheels	40 feet.
Length of bucket	12 feet 6 in.
Width of do.	2 feet.
Revolutions per minute up stream	13.5.
Diameter of cylinders	30 inches.
Length of stroke	10 feet.
Length of connecting rod	35 "
Point of cutting off steam from commencement	6 "
Number of boilers	6
Length of each boiler	30 feet.
Diameter of each boiler	42 inches.
Diameter of each flue	16 "
Grate area	98.4 sq. ft.
Diameter of each chimney	5 feet.
Height of chimneys above grates	80 "
Area over each bridge wall	42.7 sq. ft.
Area of cross section of all flues	16.7 "
Area of cross section of two chimneys	39.3 "
Heating surface of all the boilers	2617.8 "
Proportion of grate to heating surface	1 to 26.4
Proportion of grate to area bridge wall	1 to 0.47
Horse power developed by engines	1229

The fuel consumed was wood: no correct results as regards evaporation, or coal per horse power per hour, therefore given.

In conclusion, we would earnestly direct the attention of western engineers to a study of the subjects here presented, especially the Indicator and its use. We have witnessed some very faulty working engines on the Ohio, occasioned principally by the manner of working the steam; *i. e.*, the valves not being opened and shut at the proper points of the stroke. On this, everything else being in order, depends the regularity of motion and smooth working of the engine. It is true that much may be gained by practical tests; that is, by giving more or less lead to the steam valves, and by closing the exhaust valves sooner or later to give more or less cushion for the pistons to bring up against, until the engines are found to perform best; but nothing accurate can be arrived at without the application of the Indicator to every cylinder. It is therefore highly important that every engineer having charge should understand that instrument and its use. The diagrams will at first sight doubtless appear intricate and difficult to comprehend, by many of those considering themselves entirely practical; but a little study of chapter 2 of this work, together with a few applications of the instrument, and some perseverance, will soon overcome all difficulties, and result in a clear understanding of the subject, and a high appreciation of its importance.

CHAPTER VII.

BOILERS, ETC.

Boilers being the source from which the power to actuate steam engines is derived, it becomes of the first importance that not only the best and most improved types be used, but also that the proportions be such as to secure the highest results.

Since the introduction of steam to sea and river navigation, many varieties of boilers have been designed, tried, and abandoned, and many others having but little merit are still in use. As it is not, however, the purpose of these notes to give the history of inventions, but to assist in directing the mind of the student into a channel of reasoning for himself, we will for the present be content with mentioning only a few of those now most generally used; namely, the Martin water tube, the horizontal fire tube, and the western river boilers.

In designing a boiler for a steam vessel, there are many elements to be considered; such as cost, proper materials, strength to bear the intended pressure, quantity of steam to be furnished in a given time, space occupied, weight, circulation of water, durability, facilities for cleaning and repairing, requisite water and steam room, heating and grate surface, area through flues, and area and height of smoke pipe.

All things being equal, that boiler producing the largest weight of steam per given weight of combustible is the best boiler; that is, evaporating the greatest number of pounds of water per pound of fuel. By combustible is meant that portion of the fuel put into the furnaces, minus the ashes, clinker, and refuse removed.

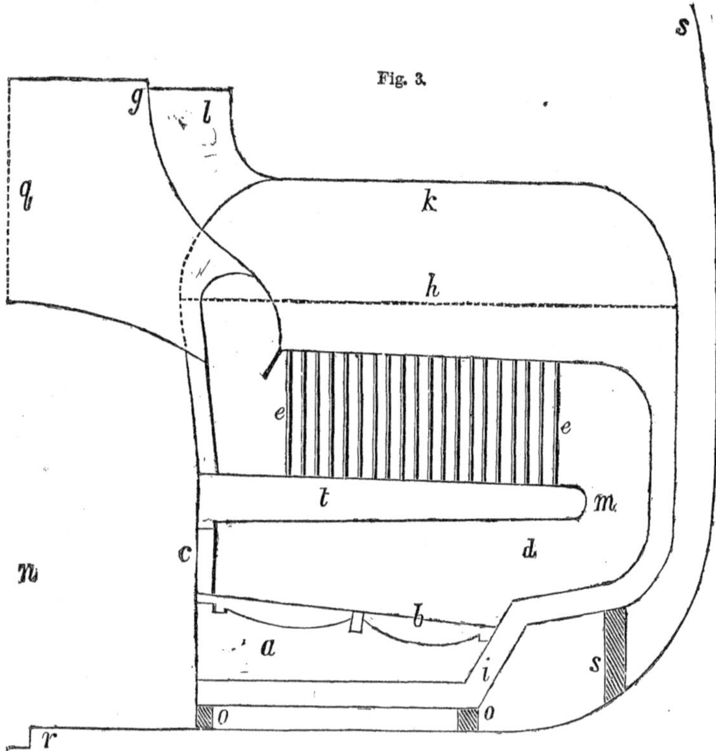

The above drawing represents a side elevation of the water-tube boiler, with the tubes arranged vertically above the furnaces, as patented by D. B. Martin, Esq., late Engineer-in-Chief of the U. S. Navy.

These boilers are almost exclusively employed in the steamers of our navy.

EXPLANATION OF THE DRAWING.

The line *r s* represents side and bottom of the ship; *o o s*, boiler keelsons, or timbers on which the boiler rests; *a*, ash pit; *c*, furnace door; *b*, grates; *d*, furnace; *m*, back connection; *e e*, the vertical tubes containing the water within them, and surrounded by the products of combustion *t*, arch over furnace; *h*, line of water level; *k*, steam room; *l*, steam chimney; *g*, passage of gases to smoke pipe; *i*, water bottom; *n*, fire-room.

These boilers are generally situated in the vessel face to face, and separated by a fire room of 8 or 9 feet, in the fore and aft direction.

The Horizontal Fire Tube, or common marine tubular boiler, has the tubes arranged horizontally above the furnaces, containing the products of combustion within, and surrounded with water. In all other respects the two types of boilers can be constructed alike. If, therefore, we imagine all the tubes to be removed from the boiler represented by the drawing, and a set of tubes arranged in it horizontally with the smoke and gases passing through them, we have the common marine tubular boiler so extensively employed in the steamers of all European nations.

To an inexperienced eye this simple difference of arrangement of tubes would doubtless appear of little or no consequence; but as simple as it may seem, it nevertheless makes an important difference in the results utilized; also in many other respects, as will be seen from the extracts given below, from a report of a Board of four Chief Engineers of the Navy, who, by the directions of the Navy Department, tested the efficiency of the two types of boilers, one of each kind having been constructed and placed on board the U. S. Steamer "San Jacinto" for the purpose of precisely the same shell, both as regards form and dimensions. The only difference between them was in the arrangement of the tubes, one being the English or horizontal fire tube; the other of the water tube type. This experiment may be considered the most important, and certainly the most extensive and accurate ever made with marine boilers.

EXTRACTS FROM REPORT.

The experiments were made to determine the relative evaporative efficiencies of the two boilers, under the conditions of actual practice on board marine steamers. For this purpose, a short experiment would be valueless from the impossibility of knowing whether the condition of the fires were exactly the same at the commencement and at the end, from the inequality in firing; from the different proportions of refuse found in different weights of coal; from fluctuations in draft; from losses by cleaning the fires; and from the different quantity of air in proportion to fuel admitted at different times. It was therefore considered necessary that the experiments with each boiler should continue uninterruptedly four days, or 96 hours.

The weight of water evaporated was ascertained from the steam pressure in the cylinders at the end of the stroke of piston, as given by the indicator. The cost of this evaporation was the weight of combustible consumed. * * * * * * Every pound of coal put into the furnace, and every pound of ashes, clinker, and refuse removed was weighed each hour. The experiments were conducted in precisely the same manner with both boilers, and as follows; namely: At the commencement, no account was taken of the coal required to raise steam, or of the temperature of the water in the boilers; but after the steam was raised to 22 lbs. per square inch pressure above the atmosphere, the level of the water in the boiler was noted, the condition of the fires estimated as nearly as possible by the eye, and the engines started. At the end of each experiment, the water in the boiler and the condition of the fires were left as at the commencement. The experiments with both boilers were begun and ended at midday, and continued uninterruptedly 96 hours. During that time, the boiler steam pressure and the vacuum in the condenser, by barometer gauges, were noted every 5 minutes; and at the close of each hour there was recorded for that hour the mean steam pressure and vacuum; the temperature of the engine room, of the fire room, of the salt and fresh water hot wells, and of the injection water; also the weight of coal thrown into the furnaces, and the weight of dry refuse in ashes, clinkers, and fine coal withdrawn. Every hour an indicator double diagram was taken from both cylinders, and from the mean of the final pressures as given by these diagrams the evaporation was calculated. * * * * * * At the commencement of each experiment, the boiler was filled with sea water; and at the expiration of every hour the saturation was recorded; also the number of inches in depth of water blown off to maintain it at 1½ times the natural concentration.

The number of double strokes made by the pistons were taken by a self-registering counter.

The same firemen fired both boilers, and the same engineers directed them. The experiments were first made on the Horizontal Fire-Tube Boilers; they were begun at noon on the 10th of June, 1859; and after being completed, the steam was shut off from it and let on from the Vertical Water-Tube Boiler, without stopping the engines. The coal was Pennsylvania anthracite.

RESULTS OF THE EXPERIMENTS.

	English Horizontal Fire-Tube Boiler.	Martin's Vertical Water-Tube Boiler.
Total number of lbs. of coal consumed	100436.00	92512.00
" " of refuse ashes, etc.	24908.00	24178.00
" " of combustible consumed	75528.00	68334.00
Per centum of coal in refuse	24.80	26.14
Mean gross horses power developed by the engines	187.25	201.07
Mean number of lbs. of coal consumed per hour	1046.21	963.67
Mean No. of lbs. of coal consumed per sq.ft. of grate	9.7	9.00
Total No. of lbs. of water evaporated from feed water of 100° Fah.	.671813455	.720 5914
Pounds of water evaporated from feed water temperature of 100° Fah. by 1 lb. of coal	6.7	7.8
Pounds of water evaporated from feed water of 100° Fah. by 1 lb. of combustible	8.9	10.6

COMPARATIVE ADVANTAGES AND DISADVANTAGES.—We are directed by your order to report to the Department the relative advantages and disadvantages of the two kinds of boiler as regards space occupied, weight, cost, accessibility for cleaning and repairs, durability, evaporative efficiency, and the relative quantities of steam that can be furnished in equal times.

1st: *As regards space.*—In the particular specimens

experimented on, the space occupied by both types of boiler was equal, but not so the area of contained heating surface. If the proper measure of that surface be, as we think it is, the extent exposed to the reception of heat from the products of combustion, then the heating surface in the vertical water-tube boiler exceeded that in the horizontal fire-tube boiler by nearly 23¾ per centum of the latter. If, however, it be measured by the extent from which water is evaporated, then the superiority will still remain with the vertical water-tube boiler, but reduced to $7\frac{1}{5}$ per centum.

2d: *As regards the weight of the two Boilers.*—By referring to the table of their dimensions and weights, it will be seen that in this respect the experimental boilers were nearly equal, the horizontal fire tube having a slight advantage in lightness; but if the aggregate weight of boiler and contained water at the steaming level be compared, then the vertical water tube has a superiority of nearly 5¼ per centum over its competitor.

3d: *Cost.*—In this particular the horizontal fire-tube boiler is slightly the cheapest, but the difference is unimportant.

4th: *Accessibility for cleaning and repairs.*—For the removal of scale or any insoluble sediment on the water surfaces of the tubes, the vertical water-tube boiler has a decisive superiority from the complete and easy manner in which the entire of those surfaces can be reached by a scaling tool and cleaned mechanically. With the horizontal fire-tube boiler this operation is very tedious and difficult, and at the best is only partial. It may indeed be said that the whole of the horizontal tubes cannot be scaled without the removal of a portion of them; and from the fact of their becoming more and more coated with scale as their age increases, their evaporative efficiency will be continuously impaired to the extent of the loss of heat thus intercepted. On the other hand, the horizontal fire tubes are much more easily and completely swept of soot and deposit from the furnaces; they are also more easily plugged when leaking. Furthermore, they are only about one fourth the number of the vertical water tubes, and the liability to leakage is correspondingly lessened, but this liability is so trifling as to be of no value in a practical estimate. The remaining portions of both boilers are equally accessible for cleaning and repairs.

5th: *Durability.*—We have no data on which to base an opinion in this respect, but we believe both boilers to be about equal.

6th: *Evaporative Efficiency.*—The relative evaporative efficiency, as given by the experiments, applies rigorously only to the particular specimens of the types of boiler employed, with their peculiarities of proportion and under the conditions of the trials; under other conditions and with other proportions, the relative evaporative efficiency would doubtless be different, and in direction as determined by better or worse proportions, and by conditions more or less favorable for one kind of boiler over the other. The proportions given to both boilers in the present case, however, are such as are now generally approved in practice. With these proportions and under the actual conditions of the trials, the evaporative efficiency of the vertical water-tube boiler exceeds that of the horizontal fire tube by 18½ per centum of the evaporation of the latter, making the comparison by weight of combustible consumed; and by $16\frac{2}{5}$ per centum if the comparison be made by weight of coal consumed; the former is, of course, the proper result.

7th: *Relative Quantities of Steam that can be furnished in equal times by the two Boilers.*—In this respect the superiority remains with the horizontal fire-tube boiler, in which the combustion of the fuel can be forced to a considerably greater extent than in the vertical water-tube boiler. The additional steam, however, thus obtained will be at a greater pro-rata cost of coal, but we have no data to determine either the increased quantity or its increased cost.

Finally, in view of the much greater evaporative efficiency of the vertical water-tube boiler, and of the facility and completeness with which it may be scaled,—the two qualities of paramount importance with marine boilers,—we would express our decided opinion that its superiority over the horizontal fire-tube boiler is so strongly marked as to unquestionably entitle it to the preference.

WESTERN RIVER BOILERS.

The first steamboat constructed for the western rivers had cylindrical boilers. Since that date, many types of boilers have been made, and tried on board some of the many steamboats navigating those immense inland waters; but none of them, except those represented by the following drawings, have ever gained general favor. In consideration of this fact, of the great number constructed every year at different places on the rivers, and of the high pressure of steam used, they deserve more than a passing notice.

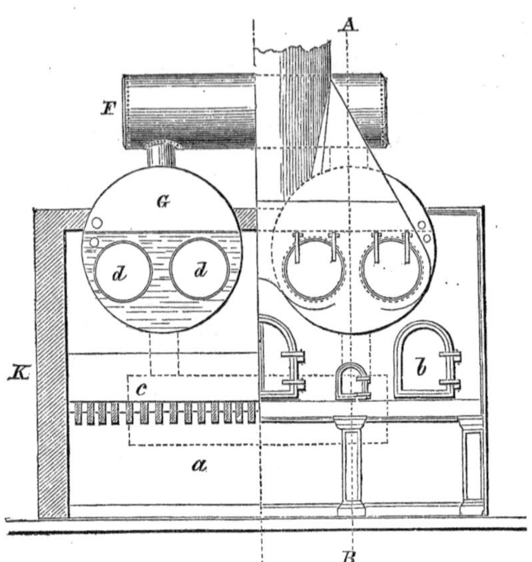

Transverse Section through Fire Bed. Front Elevation.

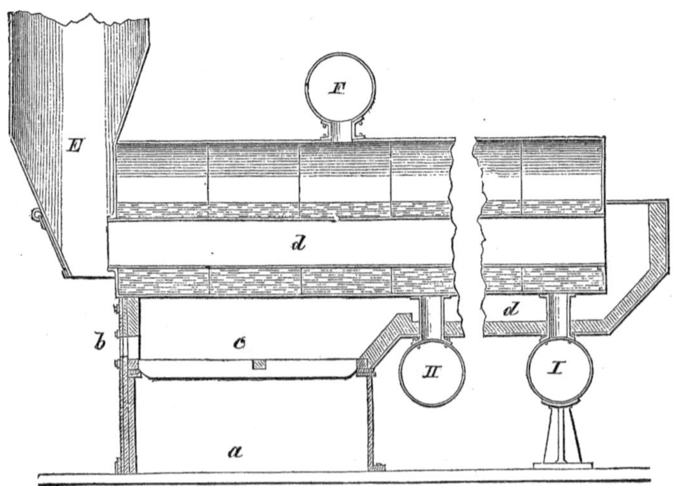

Longitudinal Section through A B of Front View.

DESCRIPTION OF DRAWINGS.

a, ash pit; *b*, furnace doors; *c*, grate bars; *d*, flues; *E*, smoke pipe; *G*, steam room; *F*, steam drum; *H*, mud blow pipe; *I*, feed; *K*, brick work.

The safety valve is put on the top of the boiler shell. These boilers are placed side by side on the deck of the vessel, near and fronting the bow, arranged in numbers from one to eight, according to the size of the boat.

The fronts are of cast iron, resting on the deck; the back ends are also supported by cast brackets under the feed. Brickwork surrounds the outside, and also forms the bottom of the lower smoke flues.

The only variations from the drawing ever made in these boilers, consists in the diameter and length of shells, and diameter and number of flues.

The majority are two-flued; but they are made with four, five, and six flues, varying in diameter from 5 to 18 inches each. The shells are made of 44, 42, 38, 36, and 34 inches diameter, with lengths varying from 32 to 22 feet. The iron used in the shells of the two largest diameters is $\frac{5}{16}$ of an inch thick, but for those of 40 inches diameter and under it is never more than $\frac{1}{4}$ inch thick; for the smallest sizes it is even occasionally $\frac{3}{16}$ thick. In the largest flues it is sometimes $\frac{5}{16}$ of an inch thick, but for all medium sizes it is $\frac{1}{4}$ inch. The heads are all wrought iron, generally $\frac{5}{8}$ of an inch thick, with the flanges turned on the front one.

All the seams are single riveted, with $\frac{5}{8}$ rivets, many of which are driven cold in all seams accessible for the purpose.

The maximum pressure of steam carried on these boilers is about 150 lbs. per square inch; but, prior to the law limiting the pressure, from 200 to 220 lbs. per square inch was not an unusual daily working pressure on a boiler of 40 inches diameter of shell, and $\frac{1}{4}$ inch thickness of iron, such as the above drawing represents; and although many explosions occurred and flues were collapsed during the early days of western river steam navigation, yet all of them have been attributed either to defective materials, imperfect workmanship, incompetent and reckless engineers, or to the omission of steam pumps (Doctors, as they are called by western engineers) to supply the boilers with water during the time that the boats were landing passengers and freight at the different stopping points.

All the iron from which they have been constructed within the past few years has been made from cold-blast charcoal pigs, worked into blooms in charcoal furnaces. The rivets are also made from the best charcoal blooms. In fact, their success as regards safety may in a great degree be attributed to the superior quality of the materials used in their construction.

The lifetime of the boats in which they are employed is averaged at five years, and when they cease to be fit for use, the engines are transferred to a new boat—sometimes to a third, and occasionally to a fourth boat: the boilers are never used on the second boat, but always removed to the shore, and worked at reduced pressures—the objections to their further use on board vessels being their reduced strength consequent upon a chemical change in the iron, occasioned by the high temperatures. Hence extreme and varied expansions and contractions; this causes crystallization, and the sheets to crack through the line of the rivet holes where the laps come directly in contact with the greatest heat. To double rivet the seams, or increase the thickness of iron, increases the evil; to cut out the defective piece and replace it with a new piece of iron starts fresh leaks, because there is a difference between the expansion of the old and new metal. The only remedy is to remove one circle of sheets.

The products of combustion are discharged into the smoke pipes at such a high temperature that it involves the necessity of making the pipes in area of cross section about twice that of the flues. The distance between the grates and bottoms of boilers does not often exceed 16 inches. The coals are frequently piled up in the furnaces to the boilers, and consequent upon the extreme height (sometimes 80 feet) of the pipes, assisted by a blast of steam discharged into the back end of the flues, or the exhaust steam admitted into the chimneys on the locomotive fashion, the draft is very strong, the combustion rapid, and the heat applied to the boiler iron intense. There are no combustion chambers or provision for admitting air through holes into the furnace doors or back of the grates. The coal is highly bituminous, used extravagantly, and as a consequence produces large volumns of dense black smoke. The cheap rate at which this coal is furnished is the only excuse for making no efforts to economize it and prevent the smoke nuisance.

We feel convinced that time must effect an entire change in the mode of generating steam on the western waters, for it is evident that tubular boilers can be constructed suitable to the purpose, that will not only be lighter and more durable, but that can be operated with 50 per cent. less fuel than those now in such high favor.

BOILER FLUES.

The well established law that the strength of cylinders is inversely as their diameters, and the hitherto undisputed axiom among practical engineers that cylindrical tubes or boiler flues when subjected to uniform external pressure were equally strong in every part regardless of length, led to erroneous opinions regarding the strength of boiler flues. For flues to collapse under the ordinary working pressure of steam, in what was supposed to be properly proportioned and well made boilers, was formerly not an unusual occurrence; and although many theories were advanced on the subject, it was not until the celebrated English engineer, William Fairbairn, LL.D., F. R. S., made an extensive set of experiments on the strengths of tubes of various forms, sizes, and lengths, that the hidden weak ness was revealed. These experiments were made by hydrostatic pressure, applied both externally and internally, to test the strength under ordinary conditions of practice, and they proved conclusively that the strength of flues exposed to external pressure, as ordinarily used, is *inversely as the length;* that is, a flue 30 feet long will collapse with just half the pressure of a flue 15 feet long, everything else being equal; in other words, a flue 30 feet long, which would bear a pressure of 100 lbs. per square inch, if shortened to 15 feet, or, what is the same thing in effect, if it be hooped in the middle of its length by angle or T iron, it will then bear a pressure of 200 lbs. per square inch.

If this important law had been familiar to the engineer who designed the boilers of the "Great Eastern," the disastrous accident which attended the first trial of that vessel would have been entirely avoided. It will be remembered that the smoke pipe of that ship was surrounded by a water jacket for the purpose of increasing the temperature of the feed water previous to entering the boilers. This jacket contained a column of water nearly forty feet in height, and the pipe was six feet in diameter; there was consequently a pressure of steam in the jacket which, united to the heavy column of water at the base, was sufficient to collapse the pipe, and cause the fearful accident.

Another source of weakness in lap joint riveted flues must also be noticed; namely, the deviation from a true circle common to the lap.

Although it had long been established, that a circle is the strongest possible form that can be made, and that no deviation from it can be made without reduction of strength, yet it was not previously known that a 9-inch diameter of tube was reduced in strength more than one-third by deviating from a circle only sufficient to make a lap joint, the ratio being as 7 to 10—so proved by the tests.

These facts are conclusive, in showing the necessity of adhering to the true circle for boiler flues using high pressure steam.

In consideration of this, and the reduction in strength consequent upon the rivet holes, it will be readily seen that the lap-welded flues, which are both seamless and round, have superior advantages over those now in use on the western waters.

RIVETING.

The weakest point is the measure of strength; therefore, in the construction of steam boilers, the riveted joints require close attention, and should receive the best workmanship. They are either single or double riveted, and the holes should not only be punched the proportionate distances apart, but should exactly correspond with each other, so that no reaming need be required.

In the United States there are three mechanical modes of uniting the sheets together; namely, *machine riveting*, and *hot* and *cold hand riveting*. On the seaboard, all kinds of boilers are riveted by the two first-named methods, in both of which the rivets are put in hot. West of the Alleghanies, all riveting is done by hand, and at Pittsburgh, Louisville, and other places, the rivets are driven cold in all places accessible for the purpose.

For the cold process, a superiority is claimed consequent upon the holes being well filled with the body of the rivets; that is, there can be no contraction—hence reduction in the strength and in the rivets' diameters after the workmen cease hammering on the heads. The reverse must be the case when driven hot; for, in cooling, the diameters are reduced by contraction. Moreover, none but the best quality of iron can be used in rivets driven cold; because, if the iron be inferior, it is sure to crack or split through the head, each one being tested by the heading.

For hot riveting, it is claimed that, in cooling, the rivets contract in length, drawing the sheets more closely together, thereby creating adhesion sufficient to add to the strength of the joint. Mr. Clarke, Resident Engineer of the Britannia Bridge, made some experiments to determine the value of this. Three plates were riveted together by a machine, maintaining a temperature of 900 degrees in the rivets; each outside plate had a circular hole in which the rivets fitted exactly; but in the centre one the hole was oval, or 2½ inches long for a ⅞ rivet, and the rivet was not allowed to touch either end of this hole. A strain was then put on the centre plate till it began to slide, which it did abruptly. Several trials were made, and the least result was an adhesion equal to 4½ tons with ⅞ rivets. Mr. Clarke infers from this experiment that, by judicious riveting, the adhesion may in many cases be nearly sufficient to counterbalance the weakening of the plates from punching the holes. In this particular we regard his opinion as in error; for if he had continued the strain on the plate until it parted, or the rivets broke, he would doubtless have found that the total pressure, or breaking strain, would have been 56 per cent. if single riveted, and 75 per cent. if double riveted, of the sheet, as fully tested by other experiments. Theoretically, there is a gain from adhesion in hot-riveted joints, but practically this seems to be lost by the contraction of the rivets' diameter, thus making the total or breaking pressure the same.

In our western river boilers, where the pressure of steam used is higher than in any part of the world, no difficulty has ever been experienced from the cold-riveted joints not being closely united and perfectly tight; and as regards strength compared with the hot riveted, superiority is claimed by those having cold-riveted boilers in charge.

In either mode of hand riveting the rivets can be seriously injured by too much hammering, and in any case by overheating. Due regard should be had to the temperature, and the blows of the hammer should be hard and quick, and not continued longer than necessary to form the head. Machine riveting has the advantage of forming the head at a single blow, and the rapidity with which the work can be performed must always give it preference over all other methods where it can be employed.

It is to be regretted that no extended set of experiments have ever been made in this country to determine the relative strengths of the different modes of riveting and uniting the sheets of steam boilers and other iron structures; also to test the relative value of the materials used in this country at the present day, for it must be evident that although the results of European experimenters on iron and steel are of value to us, yet they cannot be regarded as entirely applicable to our constructions, because our iron ores, the temperature of blast of smelting furnaces, and manner of working the metal through the different processes, and the fuel used, all differ in a large degree from those abroad.

In the year 1861 we constructed an excellent hydrostatic machine at the Navy Yard, Brooklyn, N. Y., for this very purpose, and were about to commence an extended and complete set of such experiments, when the war broke out, and we were relieved from that station for other duties.

The only published account of experiments on this

subject of any value, known to us, are those made in England by Mr. Fairbairn, and at Glasgow, Scotland, by David Kirkaldy, Esq.

The following is given by the former author, as exhibiting the strongest form and best proportions of such joints, as deduced from his experiments and actual practice.

Thickness of plates.	Diameter of rivet.	Length of rivet from head.		Distance from centre to centre.		Quantity of Lap in single riveted.		Quantity of Lap in double riveted.	
in. 16th.	in. Ratio.	in.	Ratio.	in.	Ratio.	in.	Ratio.	in.	Ratio.
0.19 = 3	0.38 = 2	0.88	4.5	1.25	6	1.25	6	2.10	10
0.25 = 4	0.50 = 2	1.13	4.5	1.50	6	1.50	6	2.50	10
0.31 = 5	0.63 = 2	1.38	4.5	1.63	5	1.88	6	3.15	10
0.38 = 6	0.75 = 2	1.63	4.5	1.75	5	2.00	5.5	3.33	9.2
0.50 = 8	0.81 = 1.5	2.25	4.5	2.00	4	2.25	4.5	3.75	7.5
0.63 = 10	0.94 = 1.5	2.75	4.	2.50	4	2.75	4.5	4.58	7.5
0.75 = 12	1.13 = 1.5	3.25	4.	3.00	4	3.25	4.5	5.42	7.5

SUPERHEATED STEAM

Is steam, the temperature of which is increased after it leaves the boilers. This is generally accomplished by passing it through a series of pipes exposed to the heat of the discharged gases in the chimney. (See drawing of such an apparatus in Nystrom's Pocket Book, page 259.) The advantage to be derived from the process depends on the condition in which the steam leaves the boilers. The theory is—steam generated in boilers, and being supplied to engines in operation, carries with it to the cylinders water in the vesicular state; that is, in minute globules or innumerable particles; and that if it be passed through a superheating apparatus, these small globules are expanded into steam, thus utilizing its full effect. This vesicular state in which steam is found in boilers when in active operation differs more or less according to the construction of the boilers. In those properly proportioned, with elevated steam chimneys, exposing much heating surface to the action of the steam, the quantity of water carried to the cylinders in this state is inconsiderable; hence, in boilers so constructed, the application of a superheating apparatus would not pay the extra cost and complication. When, therefore, we are informed of large gains in the consumption of fuel resulting from the application of such an apparatus to a set of boilers, it becomes necessary to examine the construction of the boilers before giving credit of the results to superheating the steam.

In some steamers (and this applies especially to many of those belonging to the English), where the boilers are very low, and but little height to the steam chimneys, gains in the consumption of fuel of from 20 to 40 per cent. have been effected by the application of the superheater; whereas the same kind of an apparatus, if applied to some of our Hudson River steamers, would not probably make an average gain of 5 per cent.

In the summer of 1854, by the invitation of E. K. Collins, Esq., we witnessed some extensive experiments on superheating steam, on board the splendid, but ill-fated U. S. Mail Steamship "Arctic." In this case, the superheating steam pipes were carried down the back connexions and through the furnaces close to the arches, connecting in the engine room to one common chamber, to which a branch pipe from the usual main steam pipe was also attached. Thus the arrangements were perfected to work the steam: 1st, in the usual way, and at the ordinary pressure and temperature; 2d, in a highly heated condition, from the pipes passing through the furnaces; 3d, at a medium temperature, that is, the highly heated steam mixed in the chamber with the ordinary working steam. So that a fair opportunity was offered, on a large scale, to test the benefit to be derived from superheating the steam to various temperatures; and the results of the trial proved conclusively that the saving of fuel would not pay the expense of the apparatus on board that vessel, to say nothing of the labor of keeping it in order.

This vesicular state in which steam is found in boilers must not be confounded with foaming or priming, explained at page 81.

DRAFT.

Draft is produced by the difference of temperature between a column of heated, consequently rarefied air, in a chimney, and the surrounding atmosphere; and this draft will be strong and effective just in proportion to the difference of temperatures and the height of the column. The cold and heavy column outside the chimney presses down, and forces up the warm and light column within. The greater the difference between the weight of these two columns the greater will be the draft. A column of two feet high rises, or is pressed up, with twice as much force as a column of one foot, and so in proportion for all heights to a certain extent, just as two or more corks strung together and immersed in water tend upward with proportionally more force than a single cork. In a chimney where a column of hot air one foot in height is one ounce lighter than the same bulk of external atmosphere, if the chimney be fifty feet high, the air and smoke in it is propelled upward with a power of fifty ounces.

In a correctly proportioned chimney, the area of cross section of the flues—smoke passage—should gradually contract from bottom to top, being the widest at the bottom and smallest at the top; because at the base the hot column of expanded air and gases fills the entire passage, but in ascending they gradually cool and contract, occupying less space.

There is a limit to the useful height of a chimney, consequent upon the friction of the heated air and products of combustion, and this limit is dependent on the temperature of the discharged gases compared to the area of cross section of the chimney.

APPENDIX.

MATERIALS.

There is no subject connected with the Engineering profession more important to be understood than materials. Yet young engineers rarely give it study, and but few of those in the higher grades have a thorough knowledge of the materials used in the construction of steam machinery; indeed, many cannot distinguish the best from the inferior quality of pig metal, composition of copper and tin from copper and zinc, or charcoal flange boiler plate from ordinary puddle plate.

It is in a large degree owing to this fact that we have had so many break downs in our sea-going steamers, and that the weights of those parts made of wrought and cast iron are so much in excess of what is necessary for the best materials. We have, therefore, concluded to devote a short space to this branch of the profession, solely with the view of directing the mind of the young engineer to this channel as one of his studies.

Some of the following has been extracted from the Army Ordnance Manual, and all of it accords with our experience of many months among the various iron rolling mills, furnaces, and steel works of Pennsylvania.

IRON.

Of all the useful metals none plays so important and extensive a part in the steam engine as iron. It is obtained from ores, in which it generally exists in the state of an oxide combined with earthy or stony matters, and frequently with carbon, phosphorus, sulphur, arsenic, magnesia, manganese, &c. Iron ores are classed and named according to their combinations, as magnetic, specular, clay iron-stone, red hematite, and brown hematites—the last named is the ore from which the Salisbury and Juniatta irons are extracted, the first from which the Swedish iron is obtained, and the clay iron-stone that from which the greater portion of English iron is made. For producing some varieties of pig, different kinds of ores are mixed. The foreign substances which iron is found to contain modify, in a marked degree, its essential properties. Carbon adds to its hardness, but destroys some of its characteristic qualities, and produces cast iron or steel, according to the proportion of carbon it contains; sulphur renders it fusible, difficult to weld, and brittle when heated—*hot short.* Phosphorus renders it *cold short*, but may be present in very small proportion without effecting injuriously its tenacity.

CAST IRON.

The process of making cast iron depends much on the kind of fuel used—charcoal, coke, bituminous and anthracite coals are all used, and the quality of pig metal is influenced to a great extent by the kind of fuel, as well as by the temperature of the blast with which the ore is reduced. When anthracite coal is employed, the ore is placed at once into the blast furnace. When charcoal is used, the ore is first *roasted*, by distributing it in alternate layers with waste coal, wood, or, sometimes, with charcoal, and the pile thus formed is ignited and burned in the open air. For the more refractory ores a kiln similar to that used for burning lime is required. The ore is rendered, by this operation, more porous and easily broken into small pieces, by which it is more readily acted upon in the smelting furnace. The small pieces would be disadvantageous in an anthracite furnace.

Smelting is the process by which the iron is separated from the refractory substances with which it is combined in the ore. It consists in raising the ore to a high heat in contact with carbon and a suitable flux in the blast or smelting furnace. The flux unites with the earthy matter of the ore, forming a glassy substance called *slag* or cinder, and the carbon unites with the oxygen of the ore, setting the iron free, which in turn unites with a portion of the carbon and forms a fusible compound, *carburet of iron*, or *cast iron*.

The melted iron and slag descend to the bottom of the furnace, the slag forming a covering to the pool of iron and protecting it from the action of the blast. As they accumulate, the slag runs off over the dam, and is a good indication, to an experienced eye, of the quality of metal the furnace is making.

The furnace is generally tapped once every twelve hours, and the metal is run out into channels formed in the sand, and is known as *pigs*.

Limestone is the flux used for most ores; clay is

sometimes required to mix with ores containing much limestone.

A larger yield from the same furnace, and a great economy in fuel, are effected by the use of a *hot blast.* The greater heat thus produced causes the iron to combine with a larger percentage of foreign substances, and the strength of the cast iron is thus injured.

Cast iron, for all purposes requiring great strength, should be smelted with the cold blast.

Pig iron, according to the proportion of carbon which it contains, is divided into *foundry iron* and *forge iron*, the latter being adapted only to conversion into malleable iron: while the former, containing the largest proportion of carbon, can be used either for casting or for making bar iron.

There are many varieties of cast iron, differing from each other by almost insensible shades; the two principal divisions are *gray* and *white*, so called from the color of the fracture when recent. Their properties are very different.

Gray iron is softer and less brittle than white iron; it is in a slight degree malleable and flexible, and is not sonorous; it can be easily drilled and turned in the lathe, and does not resist the file. It has a brilliant fracture, of a gray, or sometimes a bluish-gray color; the color is lighter as the grain becomes closer, and its hardness increases at the same time.

White iron is very brittle and sonorous; it resists the file and the chisel; the fracture presents a silvery appearance, generally fine grained and compact.

Mottled iron is a mixture of white and gray; it has a spotted appearance.

The pig metal is generally known in our market as

charcoal cold blast, charcoal hot blast, anthracite, and coke iron, and the quality is decided on by breaking the pigs and examining the fractures. A medium-sized grain, bright gray color, lively aspect, fracture sharp to the touch, and a close compact texture, indicate a good quality of iron. A grain either very large or very small, a dull, earthy aspect, loose texture, dissimilar crystals mixed together, indicate an inferior quality.

Besides these general divisions, the manufacturers distinguish more particularly the different varieties of pig metal by numbers, according to their relative hardness.

No. 1 is the softest iron, possessing in the highest degree the qualities described as belonging to gray iron; it has not much strength, but on account of its fluidity when melted, and of its mixing advantageously with other kinds of irons, it is of great use to the founder, and commands the highest price.

No. 2 is harder, closer grained, and stronger than No. 1; it has a gray color and considerable lustre.

No. 3 is still harder—its color is gray, but inclining to white; it has considerable strength, but is principally used for mixing with other kinds of iron.

No. 4 is bright iron, No. 5 mottled, No. 6 white—which is unfit for general use by itself.

East of the Alleghany mountains, the anthracite hot-blast iron is used for all ordinary purposes, and west of the Alleghanies coke hot-blast iron is in general use.

Pig metal is improved by being remelted in an air furnace. All cast iron expands forcibly at the moment of becoming solid, and again contracts in cooling;

gray iron expands more and contracts less than other iron.

The color and texture of cast iron depend greatly on the size of the casting and the rapidity of cooling; a small casting, which cools quickly, is almost always white, and the surface of large castings partakes more of the qualities of white metal than the interior. Care should always be taken to cool them as equally as possible, and not too rapidly.

West of the Alleghanies, where bituminous coal is so plentiful and cheap, air furnaces are in general use in foundries, and the castings made from them are superior to those from the cupola furnace, as was proved at Pittsburg, Pa., by many experiments.

MALLEABLE IRON.

The manner of converting iron ore into malleable iron has undergone many changes since the seventh descendant from Adam. Tubal Cain was an "instructor of every artificer in brass and iron." It is made from the pig, in the bloomery fire, or in the puddling furnace—generally the latter. The process consists in melting the pig metal in a reverberatory furnace, where the flame is made to act directly on the metal, keeping it exposed to a great heat, and constantly stirring the mass, bringing every part of it evenly under the action of the flame, until it loses its remaining carbon. It then loses its fluidity, and is formed into a puddler's ball, weighing from 80 to 100 pounds. This is the point or connecting link between cast and malleable iron.

The operation of puddling is a most important one,

as the quality of the iron depends so much upon the skill with which it is conducted. After the puddler's ball has been formed, it is passed to a heavy strong squeezer, or steam hammer, most frequently the former. Its object is to press out, as perfectly as possible, the liquid cinder which the ball still contains; it also forms the ball into shape. It is now called a bloom, and is ready to be rolled or hammered—while yet hot it is generally passed between the rolls several times, and drawn into a bar about five inches wide and three quarters of an inch thick; this is called muck bar. The next thing is to refine it. To prepare bars for this operation, they are cut by a strong pair of shears into such lengths as are best adapted to the size of bar or sheet required. The sheared bars are then piled one on the other, according to the quantity of metal necessary to make the finished piece. They are then brought to a welding heat, in the heating furnace, and passed between the finishing rolls successively until drawn to the proper dimensions. For heavy plates, and many other forms, pieces called tops and bottoms are first rolled some three by two feet and about an inch thick, and the before named bars piled—breaking joints—between them. Sometimes these tops and bottoms are of good stock and the pile very inferior; the result is a poor quality. For better material, the iron is double refined; *i. e.*, the rough, or muck bar, as it is called, being cut to proper lengths, is piled, heated as before, rolled into bars, and again piled, heated, and rolled to the required dimensions. Another, and we consider the better process, is to hammer the heated pile into a slab of say 18 by 12 inches and 6 inches thick, heat this slab, rehammer it, then roll it to desired dimensions. By reheating, hammering or rolling, other things being equal, the iron is improved up to some five or six workings, after which, further heating causes deterioration.

The quality depends on the kind of pig used, the skill of the puddler, and the absence of deleterious substances in the furnace. For the best sheets, bars, and for converting into steel, *charcoal iron* is used exclusively, and it can be relied upon for strength and toughness with greater confidence than any other.

Bloomery.—This resembles a large forge fire, and in it are made the charcoal blooms from which the best qualities of all iron and steel are manufactured. The pig metal, after being broken into pieces of the proper size, is placed before a strong blast directly in contact with charcoal; as the metal fuses it falls into a cavity left for the purpose below the blast, when the bloomer works it into the shape of a ball, which he places again before the blast surrounded with fresh charcoal; after repeating this, the ball is ready for and put under the hammer and hammered into a bloom. The bloom is then removed to the reverberatory furnace, and heated with bituminous coal, to a welding heat, then again hammered into a slab, or rolled into rough bar. If the former, it is again heated and rolled as required. If the latter, it is cut, piled, and re-rolled into bar—then treated as before explained. Flange boiler plate is made according to both operations, and there is a difference of opinion among iron masters regarding which process produces the best quality.

Cold Short is the term given to iron that will not

stand working cold, bending, twisting, or punching very near edges, &c.

Hot Short is iron that will not work advantageously hot, but is strong when cold. Both kinds are suitable for special purposes, but for machinery neutral iron is the kind to be relied upon. By neutral is meant iron that can be worked either cold or hot—at all ordinary temperatures.

Forging.—Good iron is often injured by being unskilfully worked. Care should be taken that the iron while heating is not exposed to the air. Iron heated for any purpose, especially for welding, should be heated as rapidly as possible, in order to expose it the least possible time to the action of the air and coal; for this purpose, the strongest fuel with an abundant steady blast is necessary.

Bessemer's Patent Process for making Malleable Iron.—We have seen that the principal impurities in cast iron consist of carbon, sulphur, phosphorus, silicon, &c. These substances, Mr. Bessemer asserts, combine with the oxygen of the atmosphere at high temperatures; he therefore runs the metal from the smelting furnace into a close vessel; when this vessel is about half full, numerous small jets of atmospheric air are forced in among the fluid metal, and in sufficient quantities to produce a vivid combustion among the particles of the fluid metal. By this process an intense heat is generated without the application of any fuel, and the labor and expense of puddling are saved. The process has been progressing in England for several years, and at this date—1862—a number of iron masters are experimenting with it in this country. If successful, a cheaper and wider field will be open to the manufacturers of iron. One of the difficulties Mr. Bessemer has to contend with, is to obtain any kind of material that will stand the intense heat.

PUDDLED STEEL.

If, in the operation of puddling, the process be stopped at a particular time, determined by indications given by the metal to an experienced eye, an iron is obtained of greater hardness and strength than ordinary iron, to which the name of semi-steel, or puddled steel, has been applied. Chemicals are also used in some such furnaces.

STEEL.

Steel is a compound of iron and carbon, in which the proportion of the latter is from five to one per cent., and even less in some kinds. Steel may be distinguished from iron by its fine grain, and its susceptibility of hardening by immersing it, when hot, in cold water.

There are many varieties of steel, the principal of which are *blistered steel*, *shear steel*, and *cast steel.*

Blistered Steel is prepared by the direct combination of iron and carbon. The process is to take the best bars and plates of wrought iron and expose them in a converting furnace, for seven or eight days, to a medium temperature, in contact with powdered charcoal, so as to totally exclude the air. The bars, on being taken out, exhibit in the fracture a uniform crystalline appearance. The degree of carbonization is varied according to the purpose for which the steel is intended.

Shear Steel is generally made from blistered steel refined by piling into fagots, which are brought to a welding heat in a reverberatory furnace, hammered and rolled again into bars; this operation is repeated several times to produce the finest kind of shear steel. The name is derived from the fact that this variety of steel was used in England for shears.

Cast Steel.—For this important invention we are indebted to Benjamin Huntsman, of the village of Handsworth, near Sheffield, England, who, about the year 1740, perfected his invention, from which the civilized world has derived such vast and varied advantages. It is made by breaking *blistered steel* or cutting *bar iron* into small pieces, and melting it in combination with a small quantity of charcoal (when it is made from iron, manganese is mixed with it) in close air-tight crucibles, from which it is poured into iron moulds; the ingot is then reduced to a bar by hammering or rolling, as described under the head of malleable iron.

Cast steel is the finest kind of steel. It is known by a very fine, even, and close grain, and a silvery and homogeneous fracture; it is very brittle, and acquires extreme hardness, but is difficult to weld without the use of a flux. The other kinds of steel have a similar appearance to cast steel, but the grain is coarser and less homogeneous; they are softer, less brittle, and weld more readily.

Properties of Steel.—The best steel possesses the following characteristics: Heated to redness and plunged into cold water, it becomes hard enough to scratch glass and to resist the best files; the hardness is uniform throughout the piece; after being tempered it is not easily broken; it welds readily; it does not crack or split; it bears a very high heat, and preserves the capability of hardening after repeated working; the grain is *fine*, *even*, and homogeneous, and it receives a brilliant polish.

Hardening and Tempering.—On these operations the quality of manufactured steel in a great measure depends.

Hardening is effected by heating the steel to cherry red and plunging it into a liquid, generally cold water; the degree of hardness depends on the heat and rapidity of cooling.

Tempering.—Steel in its hardest state being too brittle for most purposes, the requisite strength and elasticity are obtained by tempering, which is performed by heating the hardened steel to a certain degree and cooling it quickly. The requisite heat is usually ascertained by the color which the surface of the steel assumes—a straw color is common for cold chisels and machinists' tools.

Case Hardening.—This operation consists in converting the surface of wrought iron into steel, by heating the iron to a cherry red, in a close vessel, in contact with carbonaceous materials, and then plunging it into cold water. Bones, leather, hoofs, and horns of animals are used for this purpose, after having been burnt or roasted, and pulverized. Soot is also frequently used.

To Test the Quality of Boiler Iron.—Bend it cold at sharp angles, and double the pieces together; heat it cherry red, and perform the same operation, and punch holes very near the edges of the sheets. If it stands these tests without cracking, it is neutral iron, and of the best quality.

To Test the Quality of Bar Iron.—Cut a notch on one side with a cold chisel, then bend the bar over the edge of an anvil at sharp angles. If the fracture exhibits long silky fibres, of a leaden gray color, cohering together, and twisting or pulling apart before breaking, it denotes tough, soft iron, easy to work and hard to break. In general, a short, blackish fibre indicates iron badly refined. A very fine close grain denotes a hard steely iron, which is apt to be cold short, but working easily when heated, and making a good weld. Numerous cracks on the edges of the bar generally indicate a hot short iron, which cracks or breaks when punched or worked at a red heat, and will not weld. Blisters, flaws, and cinder holes are caused by imperfect welding at a low heat, or by iron not being properly worked, and do not always indicate inferior quality.

To Test Iron when Hot.—Draw a piece out, bend and twist it, split it and turn back the two parts, to see if the split extends up; finally, weld it, and observe if cracks or flaws weld easily. Good iron is frequently injured by being unskilfully worked: defects caused by this may be in part remedied. If, for example, it has been injured by cold hammering, moderate annealing heat will restore it.

Steel.—To test steel, break a few bars, taken at random, make tools of them, and try them in the severest manner.

For further information on the subject of materials, we refer the reader to an excellent work called "Useful Metals and their Alloys," by Messrs. Clay, Aitken, Vospicket, and Fairbairn.

Tenacity of Materials.

Material		Tenacity	Remarks
Cast Steel		134,000 lbs.	
Bar-iron	Swedish	72,000	Experiments by Franklin Institute, on bars whose cross section was about one-fifth of a square inch.
	Salisbury, Conn.	66,000	
	Bellefonte, Pa.	58,500	
	English	56,000	
	Pittsfield, Mass.	57,000	Experiments of Maj. W. Wade, for the Ordnance Department, on pieces whose cross section was nearly 1 square inch.
Cast-iron	Pig metal	15,000	
	Good common castings	20,000	
	Specimens from gun heads	24,000 / 39,500	
Cast Steel		128,000	
Bronze—gun metal		30,000 / 42,000	
Copper, cast, (Lake Superior)		24,138	
Brass		18,000	
Copper	Wrought	34,000	
	Cast	19,000	
Tin, cast		4,800	
Zinc		3,500	
Platinum		56,000	
Silver		40,000	
Gold		30,000	
Lead		1,800	

WOODS.

Wood	Tenacity
Ash	15,800
Mahogany	11,500
Oak	11,600
White Pine	11,800
Walnut	7,700

In general, the tenacity of metals is increased by hammering and wiredrawing. The strength of Pittsfield bar iron, given in the above table, is the mean of four trials, with cylinders 1 inch long and 0.9 inch diameter. They were extended in length, before fracture, to 1.4 in., and they were reduced in diameter to 0.6 in. in the middle.

A bar of wrought iron is extended about one-hundredth part of its length for every ton of strain on a square inch.

Transverse Strength.

$S =$ the weight in pounds required to break a beam 1 in. square and 1 in. long, fixed at one end and loaded at the other; b the breadth, d the depth, and l the length, in inches, of any other beam of the same material, and W the weight which will cause it to break, neglecting the weight of the beam itself.

1. If the beam is supported at one end, and loaded at the other:

$$W = S\frac{bd^2}{l}$$

2. If the beam is supported at one end, and the load distributed over its whole length:

$$W = 2S\frac{bd^2}{l}$$

3. If the beam is supported at both ends, and loaded in the middle:

$$W = 4S\frac{bd^2}{l}$$

4. If the beam is supported at both ends, and loaded uniformly over its whole length:

$$W = 8S\frac{bd^2}{l}$$

5. If the beam is supported at both ends, and loaded at the distance m from one end:

$$W = S\frac{lbd^2}{m(l-m)}$$

Resistance to Torsion.

S = the weight in pounds required to break, by twisting, a solid cylinder, 1 inch diameter; the weight acting at the distance of 1 inch from the axis of the cylinder; d, the diameter in inches of any other cylinder of the same material; r, the distance from its axis to the point where the breaking weight W is applied; then:

$$W = S\frac{d^3}{r}$$

Results of Repeated Heating Bar Iron.

In a series of experiments, with regard to the improvements and deterioration which result from oft-repeated heating and laminating of bar iron, made by William Clay, Esq., of the Mersey steel and iron works, Liverpool, he says that, taking a quantity of ordinary fibrous puddled iron, and reserving samples marked No. 1, we piled a portion five high, heated and rolled the remainder into bars marked No. 2, again reserving two samples from the centres of these bars, the remainder were piled as before, and so continued until a portion of the iron had undergone twelve workings.

"The following table shows the tensile strain which each number bore:

No.		Pounds.
1.	Puddled bar	43,904
2.	Re-heated	52,864
3.	"	59,585
4.	"	59,585
5.	"	57,344
6.	"	61,824
7.	"	59,585
8.	"	57,344
9.	"	57,344
10.	"	54,104
11.	"	51,968
12.	"	43,904

"It will thus be seen that the quality of the iron increased up to No. 6, (the slight difference of No. 5 may, perhaps, be attributed to the sample being slightly defective); and that from No. 6 the descent was in a similar ratio to the previous increase."

TENSILE STRENGTH OF IRON AND STEEL BARS PER SQUARE INCH.

Description of Iron and Steel.	Tensile Strength.	Authority.
Russian Iron	62,644	
English Rolled Iron	56,532	American Board of
Lawmoor "	56,103	Ordnance.
American Hammered	53,913	
Krupp's Cast Steel, average of 3 samples	111,707	Min. of War, Berlin.
Cast Steel, highest	142,222	Mallett.
" mean	88,657	do.
" "	134,256	
" tempered	150,000	
Shear Steel	124,400	
Blister "	133,152	
Mersey Steel and Iron Co. Puddled steel, highest	173,817	
Dito, another sample	160,832	
Average of three samples tested at the Liverpool Corporation testing machine	112,000	

On the strength of the joints of single and double riveted boiler plates, by William Fairbairn, Esq., F. R. S.

On comparing the strength of plates with their riveted joints, it will be necessary to examine the sectional areas taken in a line through the rivet-holes with the section of the plates themselves. It is perfectly obvious, that in perforating a line of holes along the edge of a plate, we must reduce its strength: it is also clear that the plate so perforated will be to the plate itself, nearly as the areas of their respective sections, with a small deduction for the irregularities of the pressure of the rivets upon the plate; or, in other words, the joint will be reduced in strength somewhat more than in the ratio of its section through that line to the solid section of the plate. It is evident that the rivets cannot add to the strength of the plates, their object being to keep the two surfaces of the lap in contact.

When this great deterioration of strength at the joint is taken into account, it cannot but be of the greatest importance that in structures subjected to such violent strains as boilers and ships, the strongest method of riveting should be adopted. To ascertain this, a long series of experiments were undertaken by Mr. Fairbairn, some of the results of which will be of interest here. The joint ordinarily employed in ship building is the lap joint, shown in Figs. 1 and 2. The plates to be united are made to overlap, and the rivets are passed through them, no covering-plates being required, except at the ends of the plate, where they butt against each

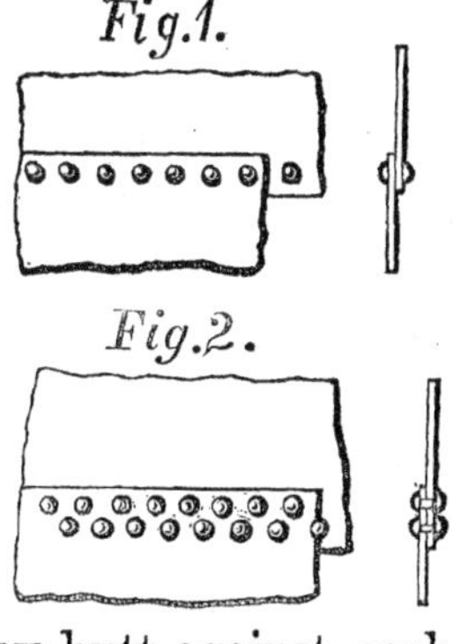

other. It is also a common practice to countersink the rivet-heads on the exterior of the vessel, that the hull may present a smooth surface for her passage through the water. This system of riveting is only used when smooth surfaces are required; under other circumstances, their introduction would not be desirable, as they do not add to the strength of the joint, but, to a certain extent, reduce it. There are two kinds of lap-joints, those said to be single-riveted (Fig. 1), and those which are double-riveted (Fig. 2). At first, the former were almost universally employed, but the greater strength of the latter has since led to their general adoption in the larger descriptions of vessels. The reason of the superiority is evident. A riveted joint gives way either by shearing off the rivets in the middle of their length, or by tearing through one of the plates in the line of the rivets. In a perfect joint, the rivets should be on the point of shearing just as the plates were about to tear; but in practice, the rivets are usually made slightly too strong. Hence, it is an established rule, to employ a certain number of rivets per lineal foot. If these are placed in a single row, the rivet-holes so nearly approach each other, that the strength of the plates is much reduced; but if they are arranged in two lines, a greater number may be used, and yet more space left between the holes, and greater strength and stiffness imparted to the plates at the joint.

The experiments of Mr. Fairbairn and others have established the following relative strengths as the value of plates with their riveted joints:

Taking the strength of the plate at..................................	100
The strength of the double-riveted joint would then be........	70
And the strength of the single-riveted joint.......................	56

THE ELEMENTS OF MACHINERY.

In consequence of having found many young engineers unacquainted with the principles of mechanical powers, we have thought best to devote a short space to the subject, prefacing it with the description of motion, and application of power, by David A. Wells, A. M.

Motion.

Motion is the act of changing place. It is absolute or relative. Absolute motion is a change of position in space, considered without reference to any other body. Relative motion is motion considered in relation to some other body, which is either in motion or at rest.

When a body commences to move from a state of rest, there must be some force to cause its motion, and this force is generally termed "Power." On the contrary, a force acting to retard a moving body, destroy its motion, or drive it in a contrary direction, is termed "Resistance." The chief causes which tend to retard or destroy the motion of a body are gravitation, friction, and resistance of the air.

The speed, or rate, at which a body moves, is termed velocity. The momentum of a body is its quantity of motion, and this expresses the force with which one body in motion would strike against another. This momentum, or force, which a moving body exerts, is estimated by multiplying its weight by its velocity. Thus a body weighing 20 lbs., and moving with a velocity of 200 feet per second, will have a momentum of $20 \times 200 = 4000$.

Action and Reaction.

When a body communicates motion to another body, it loses as much of its own momentum, or force, as it gives to the other body. The term Action is applied to designate the power which a body in motion has to impart motion, or force, to another body; and the term Reaction to express the power which the body acted upon has of depriving the acting body of its force or motion. There is no motion, or action without a corresponding and opposite action of equal amount; or, in other words, action and reaction are always equal and opposed to each other.

Application of Power.

The principal agents from whence we obtain power for practical purposes, are men and animals, water, wind, steam, and gunpowder.

When work is performed by any agent, there is always a certain weight moved over a certain space, or resistance overcome; the amount of work performed, therefore, will depend on the weight, or resistance that is moved, and the space over which it is moved. For comparing different quantities of work done by any force, it is necessary to have some standard; and this standard is the power, or labor, expended in raising a pound weight one foot high, in opposition to gravity.

A machine is an instrument, or apparatus, adapted to receive, distribute, and apply motion derived from some external force in such a way as to produce a desired result; but it cannot, under any conditions, create power, or increase the quantity of power or force applied to it. Perpetual motion, or the construction of machines which shall produce power sufficient to keep themselves in motion continually, is, therefore, an impossibility, since no combination of machinery can create, or increase, the quantity of power applied, or even preserve it without diminution.

The great general advantage that we obtain from machinery is, that it enables us to exchange time and space for power. Thus, if a man could raise to a certain height 200 pounds in one minute, with the utmost exertion of his strength, no arrangement of machinery could enable him unaided to raise 2000 pounds in the same time. If he desired to elevate this weight, he would be obliged to divide it into ten equal parts, and raise each part separately, consuming ten times the time required for lifting 200 pounds. The application of machinery would enable him to raise the whole mass at once, but would not decrease the time occupied in doing it, which would still be ten minutes.

The power will overcome the resistance of the weight, and motion will take place in a machine, when the product arising from the power multiplied by the space through which it moves in a vertical direction, is greater than the product arising from the weight multiplied by the space through which it moves in a vertical direction. Thus if a small power acts against a great resistance, the motion of the latter will be just as much slower than that of the power, as the resistance or weight is greater than the power, or if one pound be required to overcome the resistance of two pounds, the one pound must move over two feet in the same time that the resistance, two pounds, requires to move over one.

All machines, no matter how complex and intricate their construction, may be reduced to one or more of six simple machines, or elements, which we call the

Mechanical Powers.

The simple machines, six in number, are usually denominated the lever, inclined plane, wheel and axle, pulley, screw, and wedge.

The wheel and axle is, however, a revolving lever, the screw a revolving inclined plane, and the wedge a double inclined plane, thus reducing them to three in number, viz.: lever, inclined plane, and pulley.

All these machines act on the same fundamental principle of virtual velocities; accordingly, *the weight multiplied into the space it moves through is equal to the power multiplied into the space it moves through.* This is the general law which determines the equilibrium of all machines; and keeping this principle in mind, there will be no difficulty in solving any of the propositions appertaining to the simple machines.

In all machines, a portion of the effect is lost in overcoming the friction of the working parts; but, in making calculations upon them, it is made first as though no friction existed, a deduction being afterwards made. And so also we have to assume a perfection in the machine itself which does not exist; that is to say, the inclined plane, screw, wedge, &c., to be a perfectly smooth hard inflexible substance, and the rope of the pulley, and wheel and axle, to be perfectly flexible and non-elastic, conditions, for which allowance has to be made after the calculation is completed.

Lever.—Of the lever there are three orders, as shown respectively by the figures 1, 2, 3.

Fig. 1

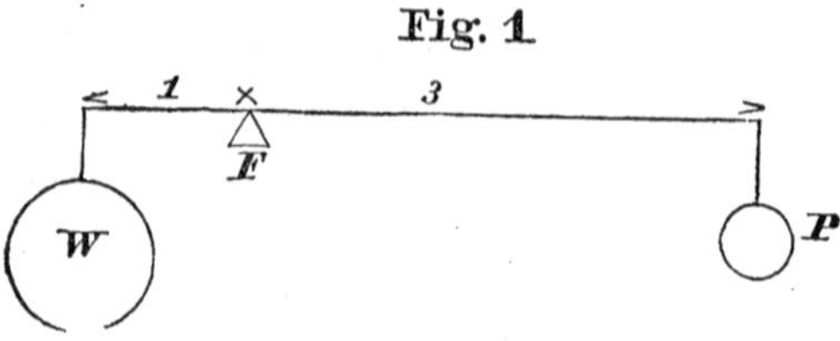

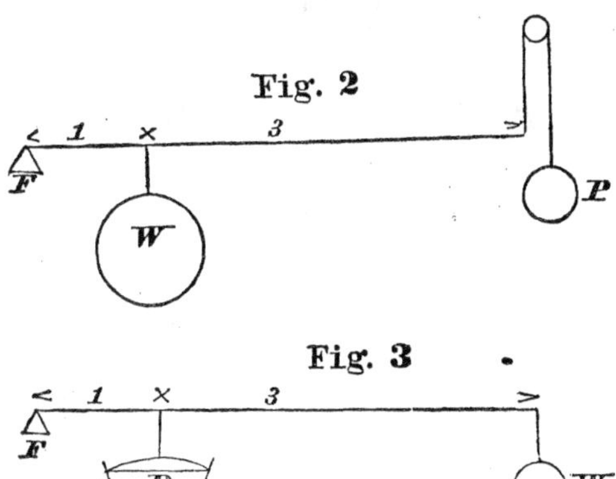

W = weight, P = power, F = fulcrum.

EXAMPLE 1.—Given the Weight W = 1000 lbs., required the power P, the lengths of the arms respectively as marked in the figures?

ANS. 1.—$P \times 3 = 1000 \times 1$
$3P = 1000$
$P = 333\frac{1}{3}$ lbs.

ANS. 2.—$P \times 4 = 1000 \times 1$
$4P = 1000$
$P = 250$ lbs.

ANS. 3.—$P \times 1 = 1000 \times 4$
$P = 4000$ lbs.

EXAMPLE 2.—Given a compound lever with lengths

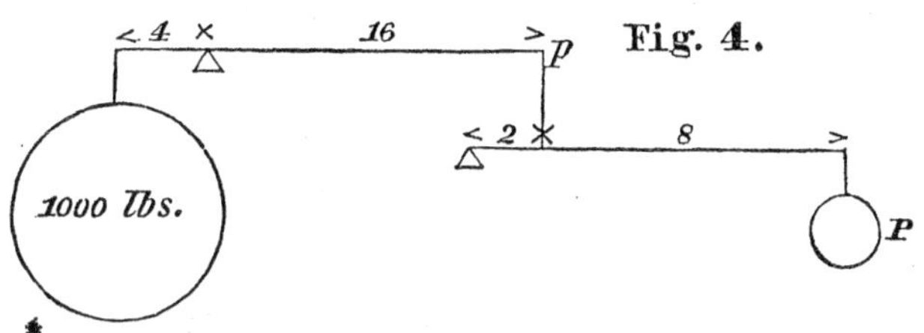

and weight as marked in fig. 4, required the power P.

$$p \times 16 = 1000 \times 4$$
$$16p = 4000$$
$p = 250$ lbs. = weight required at p, supposing there to be but one lever—therefore

$$P \times 10 = 250 \times 2$$
$$10P = 500$$
$$P = 50 \text{ lbs.}$$

Or,

$$1000 \times 4 \times 2 = P \times 10 \times 16$$
$$8000 = 160P$$
$$P = 50$$

EXAMPLE 3.—Given, as per figure 5, a safety valve

Fig. 5.

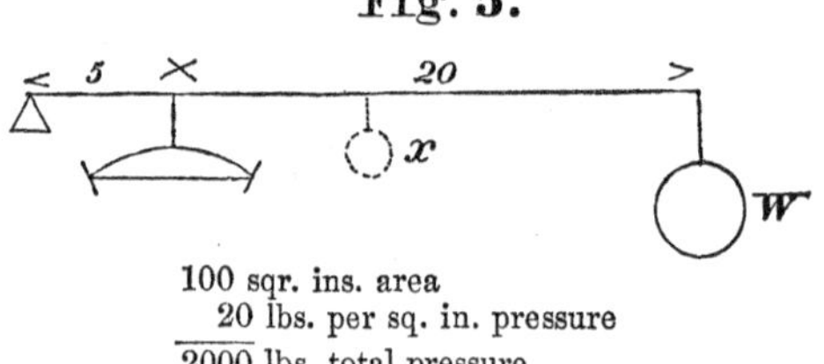

100 sq. ins. area, subject to a pressure per square inch above the atmosphere of 20 lbs., lengths of the long and short arms of the lever as shown in the figure, required the weight W to balance the pressure on the valve?

$$W \times 25 = 100 \times 20 \times 5$$
$$25W = 10000$$
$$W = 400 \text{ lbs.}$$

EXAMPLE 4.—Suppose, in example 3, the valve and stem should weigh 20 lbs., and the lever, which is uniform throughout its length, weigh 25 lbs., what would be the weight W, in that case, to balance the same pressure of steam?

The valve and stem being 5 inches from the fulcrum, act with a leverage of 5 inches, but the lever being uniform, its action is the same as though the whole weight was concentrated at x (the centre of gravity) half way of its length. Wherefore

$$W \times 25 + \overline{20 \times 5} + \overline{25 \times 12.5} = 100 \times 20 \times 5$$
$$25W + 100 + 312.5 = 10000$$
$$25W = 10000 - 412.5$$
$$W = 383.5 \text{ lbs. the re-}$$

quired weight.

Practically, the pressure a safety valve lever exerts on the valve can be ascertained by fixing it in its place, and attaching a spring balance to the pin hole immediately over the valve. If the valve and weight be also attached, the balance will indicate the total pressure which tends to keep the valve in its seat, which pressure being divided by the number of square inches in the valve, will give the pressure per square inch at which steam will commence to blow off.

Inclined Plane.

Ex. 1.—Weight W 500 lbs., length, and height of the plane, as per figure 6, 20 and 9 ins. respectively, required the power P?

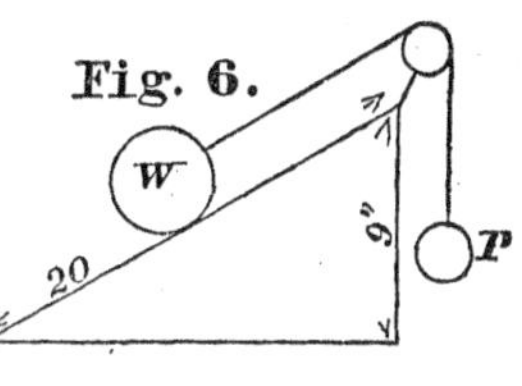

Considering the weight W to be started at the base of the plane and rolled up to the top, it will travel vertically the height of the plane, (9 inches), while the power, P, will descend a distance equal to the length of the plane (20 ins.), therefore, according to the principle of virtual velocities,

$$P \times 20 = 500 \times 9$$
$$20P = 4500$$
$$P = 225 \text{ lbs.}$$

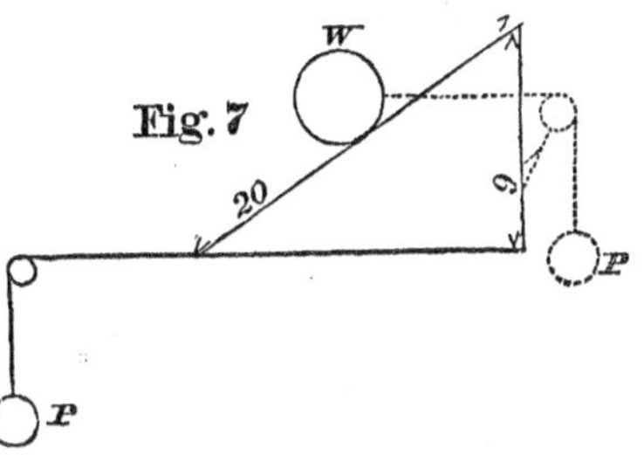

Ex. 2.—Length and height of the plane as per fig. 7, weight 500 pounds, required the

power P applied in a line with the base of the plane?

In this case, when the weight will have risen from the base to the top of the plane, 9 ins., the distance descended by P will manifestly not be equal to the length but to the base. Wherefore

$$P \times \sqrt{20^2 - 9^2} = 500 \times 9$$
$$17.86P = 4500$$
$$P = 251.96 - \text{lbs.}$$

In order to establish Equilibrium between the weight and power, this calculation is also applicable when the power is applied in the direction of the base as shown in dots, figure 7.

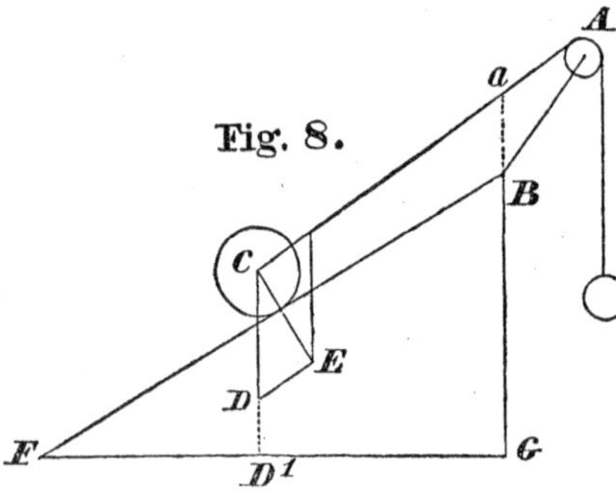

Fig. 8.

If the power be applied at an angle with the plane, as C A, figure 8, in order to ascertain the proportion of weight to the power, to establish equilibrium, we proceed thus: Draw CD, the vertical of the centre of gravity of the weight, of any convenient length; CE, at right angles to BF, and DE parallel to AC. CD can represent the length of the plane, and DE the height. Wherefore

$$\text{Weight} \times \text{DE} = \text{Power} \times \text{CD}$$
$$\text{Power} = \frac{\text{Weight} \times \text{DE}}{\text{CD}}$$

Geometrically, the angles BaC and CDE, from the construction of the figure, can be demonstrated to be equal, and also ECD, and BFG; from which, knowing the lengths of two legs of the triangle BFG, and the angle G, to be a right angle, the lengths of the lines CD ED can be determined.

Wheel and Axle.—In the wheel and axle, when the power is applied tangentially to the wheel,

W × radius of axle = P × radius of wheel
W × diameter of axle = P × diameter of wheel
W × circumference of axle = P × circum. of wheel.

When the power is not applied tangentially to the wheel, but in the direction shown in fig. 9, the length of the line *ab* at right angles to the power will give the leverage of the power,—hence

Fig. 9.

W × radius of axle = P × *ab*.

Pulley.—If a cord be pulled at one end the tension throughout its whole length must be alike. Taking figure 10, and supposing the power to be 1, the tension throughout the entire length of the cord will be 1, but as there are two parts of the cord supporting the lower block, the weight must be 2. The pressure on the fulcrum or support must be always equal to the weight, plus the power. If there be more than one support, the sum of the pressures on them will be equal to the sum of the weight and power. Or, in figure 10, according to the principle of virtual velocities, the weight is double the power, because the power must descend 2 feet for every foot ascent of the weight.

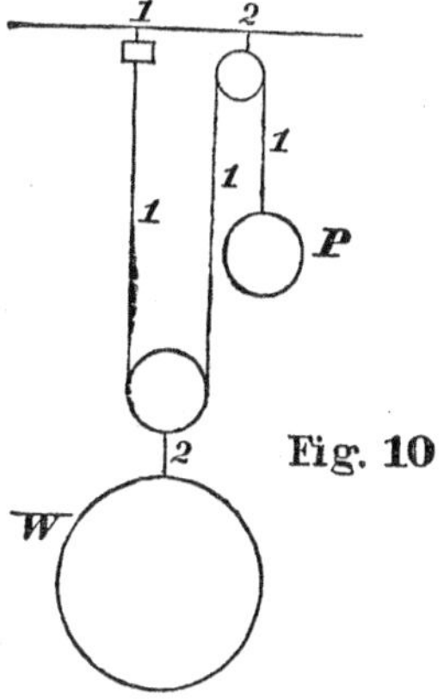

Fig. 10

The numbers above the top blocks in all the examples of pulleys here shown represent the pressure on the supports.

In fig. 11, the power and weight are as 1 to 8, because the power supports 4 weights. each one double its size.

Fig. 11

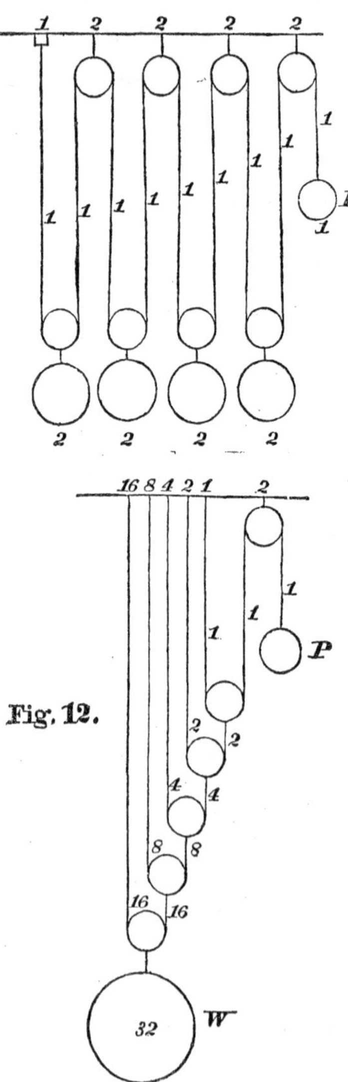

Fig. 12.

In fig. 12 the tension on the 1st cord is 1; on the 2d 2; 3d 4; 4th 8; 5th 16; and as there are 2 parts of the cord having a tension of 16, the weight to establish equilibrium, must be 32.

In fig. 13 the weight to the power is as 3 to 1, there being 3 parts of the cord having a tension of 1 supporting the weight.

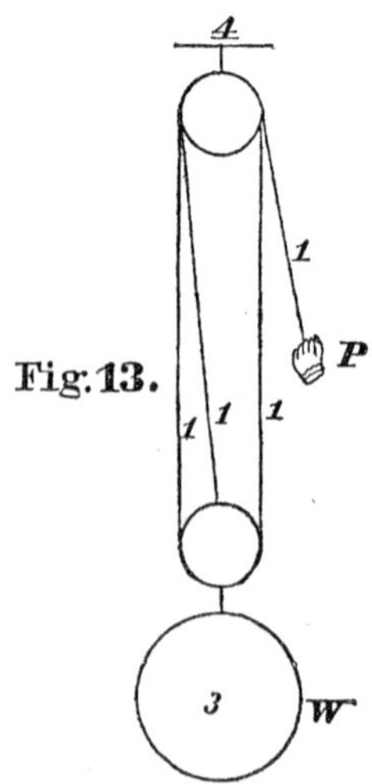

Fig. 13.

In fig. 14 the power to the weight is as 1 to 12, the power being multiplied four times by the application of the second set of pulleys, or luff-tackles, as they are technically termed.

In fig. 15 the power is to the weight as 1 to 12, the tension throughout the first cord being 1; the second cord 2; third 5, and as there are two parts of the cord having a tension of 5, and one part of the cord having a tension of 2, supporting the weight, if all the cords be supposed parallel, the weight must be the sum of these, or 12.

In fig. 16 the power to the weight is as 1 to 4.

In figure 17, where the power is applied at an angle, we ascertain the proportion of the weight and power thus: Draw AD, of any convenient length, and from the point A draw AB parallel to

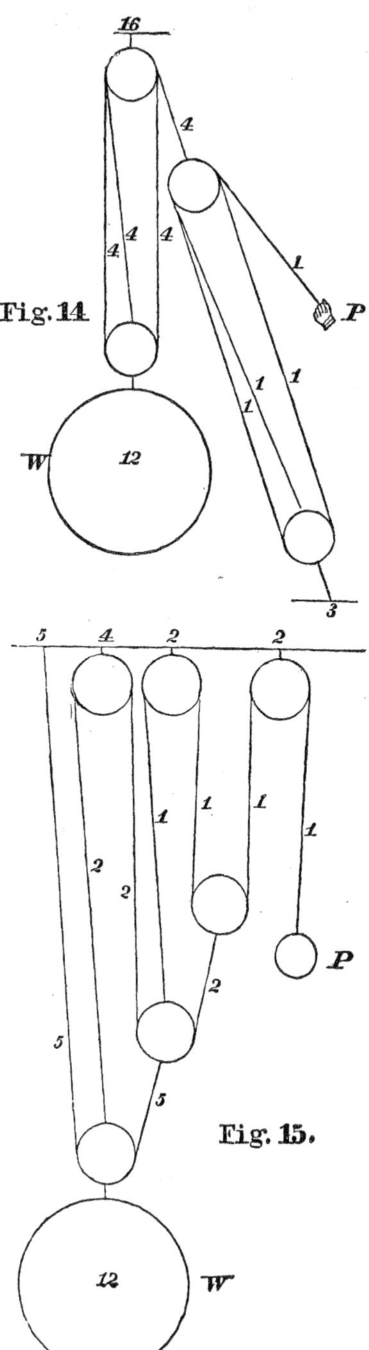

Fig. 14

Fig. 15.

Cc and AC parallel to B*b*. The power and weight will be respectively as the lengths of the lines DC or DB and AD.

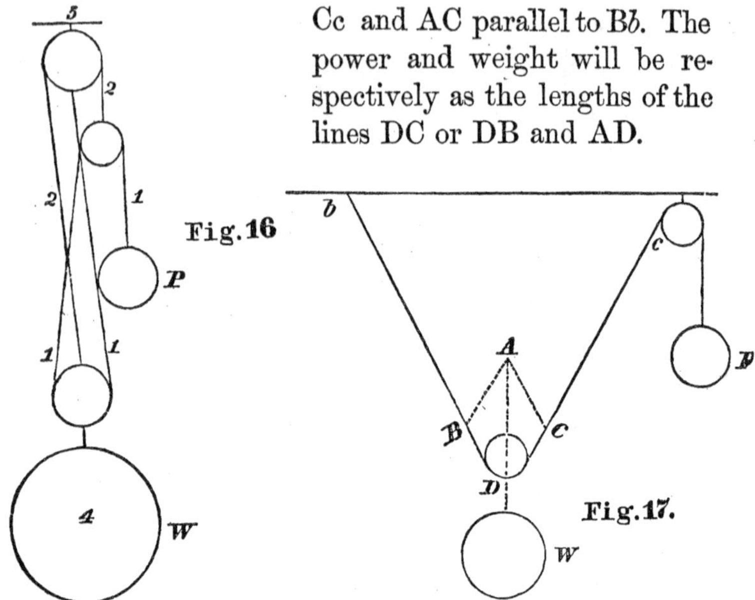

Fig. 16

Fig. 17.

From which it will be seen that the greater the angle CDB the longer will be the line DC or DB, and hence the greater the power. So that the weight of the line itself will be sufficient to prevent any powei whatever from drawing it mathematically straight.

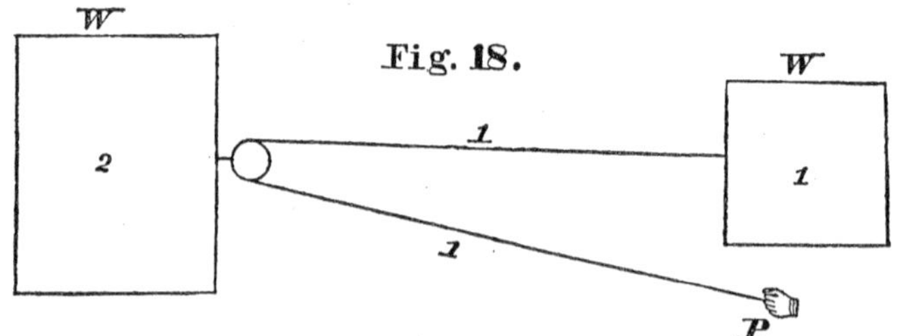

Fig. 18.

Question.—In figure 18, two blocks of granite, joined together as shown, are laid upon a horizontal plane; required their relative sizes in order that they may commence at the same time to move, and continue to move with equal velocity?

Ans.—2 to 1, because the larger block is supported by two parts ot the cord, and has in consequence, double the force exerted upon it of the smaller block.

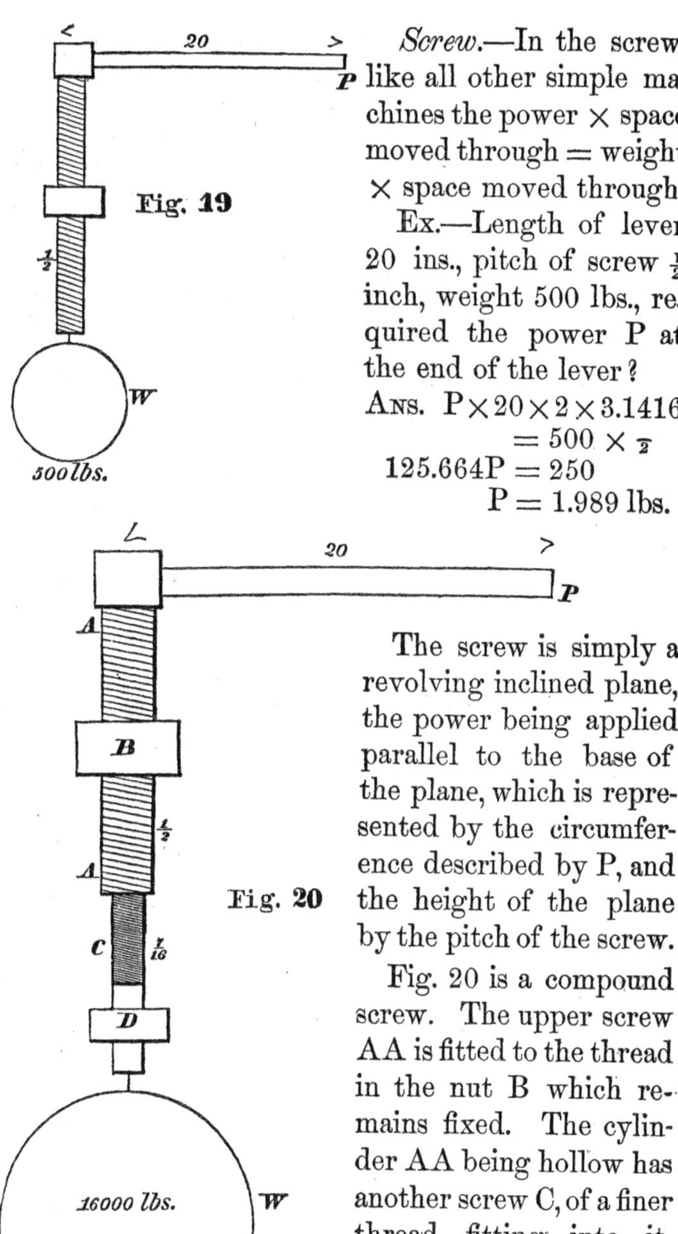

Fig. 19

Screw.—In the screw, like all other simple machines the power × space moved through = weight × space moved through.

Ex.—Length of lever 20 ins., pitch of screw ½ inch, weight 500 lbs., required the power P at the end of the lever?

ANS. $P \times 20 \times 2 \times 3.1416 = 500 \times \frac{1}{2}$

$125.664P = 250$

$P = 1.989$ lbs.

Fig. 20

The screw is simply a revolving inclined plane, the power being applied parallel to the base of the plane, which is represented by the circumference described by P, and the height of the plane by the pitch of the screw.

Fig. 20 is a compound screw. The upper screw AA is fitted to the thread in the nut B which remains fixed. The cylinder AA being hollow has another screw C, of a finer thread, fitting into it. The nut D is fixed, allowing C to slide up and down within it, without turning. By this arrangement it will be seen, that when the screw AA is turned once round, the distance ascended by the weight will not be equal to the pitch of AA, but the difference between the pitch of AA and C.

Example.—Pitch of AA $\frac{1}{2}$ inch, of C $\frac{7}{16}$ inch, weight 16000 lbs., required the power P, applied 20 inches from the centre?

$$\text{Ans.}—P \times 20 \times 2 \times 3.1416 = 16000 \times \overline{\tfrac{8}{16} - \tfrac{7}{16}}$$
$$125.664P = 1000$$
$$P = 7.957 \text{ lbs.}$$

In order to multiply the power the same number of times with a single screw, the pitch would have to be $\frac{1}{16}$ inch, which would render the thread too weak to withstand a heavy pressure.

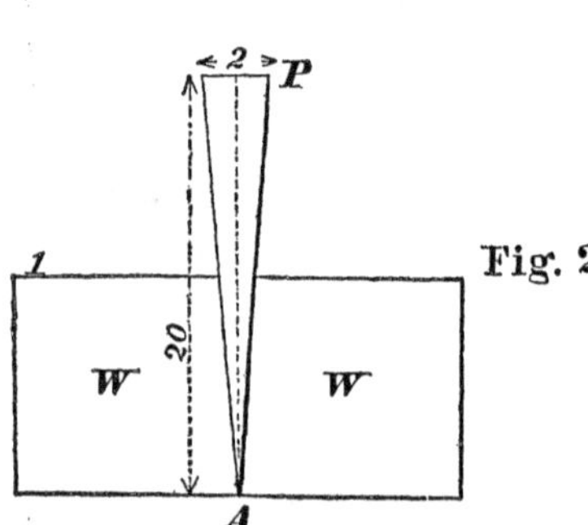

Fig. 21.

Wedge.—Let WW, fig. 21, be two weights of 1000 lbs. each, resting upon a horizontal plane, required the power to be applied at P, to the wedge, having the dimensions shown in the figure to to separate them?

$$P \times 20 = 1000 \times 2$$
$$20P = 2000$$
$$P = 100 \text{ lbs.}$$

Because, when the power P has descended to the point A, the weights have been separated 2 inches while the power has travelled 20 inches, the length of the wedge.

Centre of Gravity.

The centre of gravity of a cone from the vertex equals $\frac{3}{4}$ the axis.

In a paraboloid, the distance from vertex equals $\frac{2}{3}$ the axis.

In a parabolic space, equals $\frac{3}{5}$ the axis from the vertex.

In a triangle, equals $\frac{2}{3}$ the axis from the vertex.

Centre of Pressure.

The centre of pressure of a parallelogram, when the upper surface is level with the water, $= \frac{1}{3}$ from the bottom; of a right-angled triangle with the base down $= \frac{1}{4}$ from the bottom, measured on the perpendicular line B C; with the base up $= \frac{1}{2}$ B C.—*See Hann's Mechanics.*

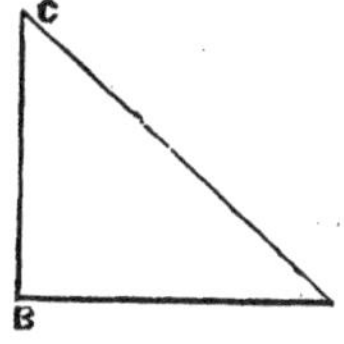

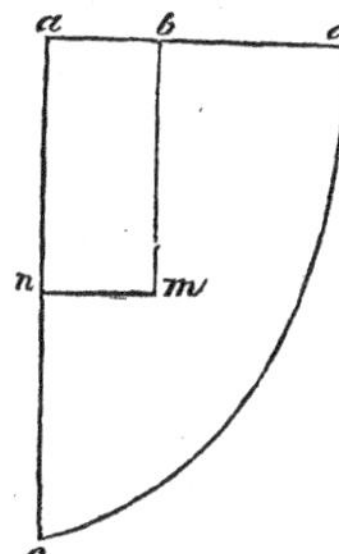

Semi-parabolic plane.

FORMULA:

$m =$ centre of pressure,

$b\ m = \frac{4}{7}$ of $a\ c$,

$m\ n = \frac{5}{16}$ of $a\ d$.

Gravity.

The spaces described by a body acted upon freely by gravity are as the squares of the times; *i. e.*, a body falling 2 seconds, will describe 4 times the distance of a body falling one second. Hence, in order to ascertain the distance fallen by a body, it is only necessary to multiply the square of the number of seconds by the distance fallen in the first second; the product will be the total distance fallen.

All bodies fall with the same velocity in vacuo, namely, 16.08 feet in the first second, having a velocity of 32.166 feet at the end of the second. Where the atmosphere is interposed, the velocity will be somewhat less, say for heavy bodies, such as the metals, 16 feet for the first second.

EXAMPLE.—Which will strike with the greater effect, a weight of 200 lbs., falling through 144 ft., or 100 lbs. falling through 256 feet?

The velocity of a body at the end of a fall is equal to the number of seconds it is falling, multiplied into (32 feet) the velocity at the end of the first second, and the *momentum* of a body is equal to the weight multiplied into the velocity. We have, then, first to find the velocity, and afterwards the momentum.

$\sqrt{16} : 1 :: \sqrt{144} : 3$ seconds time of falling of 200 lb.

$\sqrt{16} : 1 :: \sqrt{256} : 4$ " " " 100 lb.

$32 \times 3 = 96$ ft. per second velocity at end of fall of 200 lb. weight.

$32 \times 4 = 128$ ft. per second velocity at end of fall of 100 lb. weight.

$96 \times 200 = 19200 =$ momentum of the 200 lb. weight.

$128 \times 100 = 12800 =$ momentum of the 100 lb. weight.

$6400 =$ difference, which is $33\frac{1}{3}$ per cent. of the larger number.

Centre of Gravity of Several Bodies taken together.

Suppose there be several weights placed as follows in the same plane, required the centre of gravity of them all taken together?

Cylinder. Tons.		Air-pump. Tons.		Shaft. Tons.		Boilers. Tons.	
5		2		10		30	
$<$	8 ft.	$\times$	10 ft	$\times$	20 ft.	$>$	*a*

Assume a point (*a*), at any distance (say 2 feet) from either of the extreme weights, and multiply each weight separately by its distance from this point; the sum of these products, divided by the sum of the weights, will be the distance of the centre of gravity from the assumed point. Thus:

$$\begin{aligned} 30 \times 2 &= 60 \\ 10 \times 22 &= 220 \\ 2 \times 32 &= 64 \\ 5 \times 40 &= 200 \end{aligned}$$

$$47 \;)\; 544 \;(11.57 \text{ ft.}$$

= centre of gravity from the point *a*, or 9.57 feet from the boilers towards the shaft.

Displacement of Fluids.

Solid bodies immersed in fluids will displace an amount of the fluid equal to their own weight. If the specific gravity of the body be greater than that of the fluid, it will sink; otherwise it will float.

EXAMPLE.—Required the distance a cube of cherry, one foot high, will sink in fresh water?

The specific gravities of fresh water and cherry are relatively as 1.00 to .606; the cherry will therefore sink .606 feet.

Table of the Elastic Force, Temperature, and Volume of Steam, from a Temperature of 80° *to* 387.3°, *and from a Pressure of one to* 410 *Inches of Mercury.*

Elastic force in		Temperature.	Volume.	Elastic force in		Temperature.	Volume.
inches of mercury.	pounds per sq. inch.			inches of mercury.	pounds per sq. inch.		
1	.49	80	41031	53.04	26	243.3	1007
1.17	.573	85	35393	55.08	27	245.5	973
1.36	.666	90	30425	57.12	28	247.6	941
1.58	.774	95	26686	59.16	29	249.6	911
1.86	.911	100	22873	61.2	30	251.6	883
2.04	1	103	20958	63.24	31	253.6	857
2.18	1.068	105	19693	65.28	32	255.5	833
2.53	1.24	110	16667	67.32	33	257.3	810
2.92	1.431	115	14942	69.36	34	259.1	788
3.33	1.632	120	13215	71.4	35	260.9	767
3.79	1.857	125	11723	73.44	36	262.6	748
4.34	2.129	130	10328	75.48	37	264.3	729
5	2.45	135	9036	77.52	38	265.9	712
5.74	2.813	140	7938	79.56	39	267.5	695
6.53	3.1	145	7040	81.6	40	269.1	679
7.42	3.636	150	6243	83.64	41	270.6	664
8.4	4.116	155	5559	85.68	42	272.1	649
9.46	4.635	160	4976	87.72	43	273.6	635
10.68	5.23	165	4443	89.76	44	275	622
12.13	5.94	170	3943	91.8	45	276.4	610
13.62	6.67	175	3838	93.84	46	277.8	598
15.15	7.42	180	3208	95.88	47	279.2	586
17	8.33	185	2879	97.92	48	280.5	573
19	9.31	190	2595	99.96	49	281.9	564
21.22	10.4	195	2342	102	50	283.2	554
23.64	11.58	200	2118	104.04	51	284.4	544
26.13	12.7	205	1932	106.08	52	285.7	534
28.84	14.13	210	1763	108.12	53	286.9	525
29.41	14.41	211	1730	110.16	54	288.1	516
30	14.7	212	1700	112.02	55	289.3	508
30.6	15	212.8	1669	114.24	56	290.5	500
31.62	15.5	214.5	1618	116.28	57	291.7	492
32.64	16	216.3	1573	118.32	58	292.9	484
33.66	16.5	218	1530	120.36	59	294.2	477
34.68	17	219.6	1488	122.4	60	295.6	470
35.7	17.5	221.2	1440	124.44	61	296.9	463
36.72	18	222.7	1411	126.48	62	298.1	456
37.74	18.5	224.2	1377	128.52	63	299.2	449
38.76	19	225.6	1343	130.56	64	300.3	443
39.78	19.5	227.1	1312	132.62	65	301.3	437
40.80	20	228.5	1281	134.64	66	302.4	431
41.82	20.5	229.9	1253	136.68	67	303.4	425
42.84	21	231.2	1225	138.72	68	304.4	419
43.86	21.5	232.5	1199	140.76	69	305.4	414
44.88	22	233.8	1174	142.8	70	306.4	408
45.90	22.5	235.1	1150	144.84	71	307.4	403
46.92	23	236.3	1127	146.88	72	308.4	398
46.94	23.5	237.5	1105	148.92	73	309.3	393
48.96	24	238.7	1084	150.96	74	310.3	388
49.98	24.5	239.9	1064	153.02	75	311.2	383
51.	25	241	1044	155.06	76	312.2	379

Elastic force in		Tempera-ture.	Volume.	Elastic force in		Tempera-ture.	Volume.
inches of mercury.	pounds per sq. inch.			inches of mercury.	pounds per sq. in.		
157.1	77	313.1	374	254.99	125	349.1	240
159.14	78	314	370	265.19	130	352.1	233
161.18	79	314.9	366	275.39	135	355	224
163.22	80	315.8	362	285.59	140	357.9	218
165.26	81	316.7	358	295.79	145	360.6	210
167.3	82	317.6	354	306	150	363.4	205
169.34	83	318.4	350	316.19	155	366	198
171.38	84	319.3	346	326.39	160	368.7	193
173.42	85	320.1	342	336.59	165	371.1	187
183.62	90	324.3	325	346.79	170	373.6	183
193.82	95	328.2	310	357	175	376	178
203.99	100	332	295	367.2	180	378.4	174
214.19	105	335.8	282	377.1	185	380.6	169
224.39	110	339.2	271	387.6	190	382.9	166
234.59	115	342.7	259	397.8	195	384.1	161
244.79	120	345.8	251	408	200	387.3	158

Scientific Books.

FRANCIS' (J. B.) Hydraulic Experiments. Lowell Hydraulic Experiments—being a Selection from Experiments on Hydraulic Motors, on the Flow of Water over Weirs, and in Open Canals of Uniform Rectangular Section, made at Lowell, Mass. By J. B. Francis, Civil Engineer. Second edition, revised and enlarged, including many New Experiments on Gauging Water in Open Canals, and on the Flow through Submerged Orifices and Diverging Tubes. With 23 copperplates, beautifully engraved, and about 100 new pages of text. 1 vol., 4to. Cloth. $15.

Most of the practical rules given in the books on hydraulics have been determined from experiments made in other countries, with insufficient apparatus, and on such a minute scale, that in applying them to the large operations arising in practice in this country, the engineer cannot but doubt their reliable applicability. The parties controlling the great water-power furnished by the Merrimack River at Lowell, Massachusetts, felt this so keenly, that they have deemed it necessary, at great expense, to determine anew some of the most important rules for gauging the flow of large streams of water, and for this purpose have caused to be made, with great care, several series of experiments on a large scale, a selection from which are minutely detailed in this volume.

The work is divided into two parts—Part I., on hydraulic motors, includes ninety-two experiments on an improved Fourneyron Turbine Water-Wheel, of about two hundred horse-power, with rules and tables for the construction of similar motors:—Thirteen experiments on a model of a centre-vent water-wheel of the most simple design, and thirty-nine experiments on a centre vent water-wheel of about two hundred and thirty horse-power.

Part II. includes seventy-four experiments made for the purpose of determining the form of the formula for computing the flow of water over weirs; nine experiments on the effect of backwater on the flow over weirs; eighty-eight experiments made for the purpose of determining the formula for computing the flow over weirs of regular or standard forms, with several tables of comparisons of the new formula with the results obtained by former experimenters; five experiments on the flow over a dam in which the crest was of the same form as that built by the Essex Company across the Merrimack River at Lawrence, Massachusetts; twenty-one experiments on the effect of observing the depths of water on a weir at different distances from the weir; an extensive series of experiments made for the purpose of determining rules for gauging streams of water in open canals, with tables for facilitating the same; and one hundred and one experiments on the discharge of water through submerged orifices and diverging tubes, the whole being fully illustrated by twenty-three double plates engraved on copper.

In 1855 the proprietors of the Locks and Canals on Merrimack River, at whose expense most of the experiments were made, being willing that the public should share the benefits of the scientific operations promoted by them, consented to the publication of the first edition of this work, which contained a selection of the most important hydraulic experiments made at Lowell up to that time. In this second edition the principal hydraulic experiments made there, subsequent to 1855, have been added, including the important series above mentioned, for determining rules for the gauging the flow of water in open canals, and the interesting series on the flow through a submerged Venturi's tube, in which a larger flow was obtained than any we find recorded.

FRANCIS (J. B.) On the Strength of Cast-Iron Pillars, with Tables for the use of Engineers, Architects, and Builders. By James B. Francis, Civil Engineer. 1 vol., 8vo. Cloth. $2.

HOLLEY'S RAILWAY PRACTICE. American and European Railway Practice, in the Economical Generation of Steam, including the materials and construction of Coal-burning Boilers, Combustion, the Variable Blast, Vaporization, Circulation, Superheating, Supplying and Heating Feed-water, &c., and the adaptation of Wood and Coke-burning Engines to Coal-burning; and in Permanent Way, including Road-bed, Sleepers, Rails, Joint Fastenings, Street Railways, &c., &c. By ALEXANDER L. HOLLEY, B. P. With 77 lithographed plates. 1 vol., folio. Cloth. $12.

"This is an elaborate treatise by one of our ablest civil engineers, on the construction and use of locomotives, with a few chapters on the building of Railroads. * * * All these subjects are treated by the author, who is a first-class railroad engineer, in both an intelligent and intelligible manner. The facts and ideas are well arranged, and presented in a clear and simple style, accompanied by beautiful engravings, and we presume the work will be regarded as indispensable by all who are interested in a knowledge of the construction of railroads and rolling stock, or the working of locomotives."—*Scientific American.*

HENRICI (OLAUS). Skeleton Structures, especially in their Application to the Building of Steel and Iron Bridges. By OLAUS HENRICI. With folding plates and diagrams. 1 vol., 8vo. Cloth. $3.

WHILDEN (J. K.) On the Strength of Materials used in Engineering Construction. By J. K. WHILDEN. 1 vol., 12mo. Cloth. $2.

"We find in this work tables of the tensile strength of timber, metals, stones, wire, rope, hempen cable, strength of thin cylinders of cast-iron; modulus of elasticity, strength of thick cylinders, as cannon, &c., effects of reheating, &c., resistance of timber, metals, and stone to crushing; experiments on brick-work; strength of pillars; collapse of tube; experiments on punching and shearing; the transverse strength of materials; beams of uniform strength; table of coefficients of timber, stone, and iron; relative strength of weight in cast-iron, transverse strength of alloys; experiments on wrought and cast-iron beams: lattice girders, trussed cast-iron girders; deflection of beams; torsional strength and torsional elasticity."—*American Artisan.*

CAMPIN (F.) On the Construction of Iron Roofs. A Theoretical and Practical Treatise. By FRANCIS CAMPIN. With wood-cuts and plates of Roofs lately executed. Large 8vo. Cloth. $3.

BROOKLYN WATER-WORKS AND SEWERS. Containing a Descriptive Account of the Construction of the Works, and also Reports on the Brooklyn, Hartford, Belleville, and Cambridge Pumping Engines. Prepared and printed by order of the Board of Water Commissioners. With illustrations. 1 vol., folio. Cloth. $15.

ROEBLING (J. A.) Long and Short Span Railway Bridges. By JOHN A. ROEBLING, C. E. Illustrated with large copperplate engravings of plans and views. Imperial folio, cloth. $25.

CLARKE (T. C.) Description of the Iron Railway Bridge across the Mississippi River at Quincy, Illinois. By THOMAS CURTIS CLARKE, Chief Engineer. Illustrated with numerous lithographed plans. 1 vol., 4to. Cloth. $7.50.

WILLIAMSON (R. S.) On the Use of the Barometer on Surveys and Reconnaissances. Part I. Meteorology in its Connection with Hypsometry. Part II. Barometric Hypsometry. By R. S. WILLIAMSON, Bvt. Lieut.-Col. U. S. A., Major Corps of Engineers. With Illustrative Tables and Engravings. Paper No. 15, Professional Papers, Corps of Engineers. 1 vol., 4to. Cloth. $15.

"SAN FRANCISCO, CAL., *Feb.* 27, 1867.

"Gen. A. A. HUMPHREYS, Chief of Engineers, U. S. Army:

"GENERAL—I have the honor to submit to you, in the following pages, the results of my investigations in meteorology and hypsometry, made with the view of ascertaining how far the barometer can be used as a reliable instrument for determining altitudes on extended lines of survey and reconnaissances. These investigations have occupied the leisure permitted me from my professional duties during the last ten years, and I hope the results will be deemed of sufficient value to have a place assigned them among the printed professional papers of the United States Corps of Engineers. Very respectfully, your obedient servant,

"R. S. WILLIAMSON,

"Bvt. Lt.-Col. U. S. A., Major Corps of U. S. Engineers."

TUNNER (P.) A Treatise on Roll-Turning for the Manufacture of Iron. By PETER TUNNER. Translated and adapted. By JOHN B. PEARSE, of the Pennsylvania Steel Works. With numerous engravings and wood-cuts. 1 vol., 8vo., with 1 vol. folio of plates. Cloth. $10.

SHAFFNER (T. P.) Telegraph Manual. A Complete History and Description of the Semaphoric, Electric, and Magnetic Telegraphs of Europe, Asia, and Africa, with 625 illustrations. By TAL. P. SHAFFNER, of Kentucky. New edition. 1 vol., 8vo. Cloth. 850 pp. $6.50.

MINIFIE (WM.) Mechanical Drawing. A Text-Book of Geometrical Drawing for the use of Mechanics and Schools, in which the Definitions and Rules of Geometry are familiarly explained; the Practical Problems are arranged, from the most simple to the more complex, and in their description technicalities are avoided as much as possible. With illustrations for Drawing Plans, Sections, and Elevations of Buildings and Machinery; an Introduction to Isometrical Drawing, and an Essay on Linear Perspective and Shadows. Illustrated with over 200 diagrams engraved on steel. By WM MINIFIE, Architect. Seventh edition. With an Appendix on the Theory and Application of Colors. 1 vol., 8vo. Cloth. $4.

"It is the best work on Drawing that we have ever seen, and is especially a text-book of Geometrical Drawing for the use of Mechanics and Schools. No young Mechanic, such as a Machinist, Engineer, Cabinet-Maker, Millwright, or Carpenter should be without it."—*Scientific American.*

"One of the most comprehensive works of the kind ever published, and cannot but possess great value to builders. The style is at once elegant and substantial."—*Pennsylvania Inquirer.*

"Whatever is said is rendered perfectly intelligible by remarkably well-executed diagrams on steel, leaving nothing for mere vague supposition; and the addition of an introduction to isometrical drawing, linear perspective, and the projection of shadows, winding up with a useful index to technical terms."—*Glasgow Mechanics' Journal.*

☞ The British Government has authorized the use of this book in their schools of art at Somerset House, London, and throughout the kingdom.

MINIFIE (WM.) Geometrical Drawing. Abridged from the octavo edition, for the use of Schools. Illustrated with 48 steel plates. Fifth edition, 1 vol., 12mo. Half roan. $1.50.

"It is well adapted as a text-book of drawing to be used in our High Schools and Academies where this useful branch of the fine arts has been hitherto too much neglected."—*Boston Journal.*

PEIRCE'S SYSTEM OF ANALYTIC MECHANICS. Physical and Celestial Mechanics, by Benjamin Peirce, Perkins Professor of Astronomy and Mathematics in Harvard University, and Consulting Astronomer of the American Ephemeris and Nautical Almanac. Developed in four systems of Analytic Mechanics, Celestial Mechanics, Potential Physics, and Analytic Morphology. 1 vol., 4to. Cloth. $10.

GILLMORE. Practical Treatise on Limes, Hydraulic Cements, and Mortars. Papers on Practical Engineering, U. S. Engineer Department, No. 9, containing Reports of numerous experiments conducted in New York City, during the years 1858 to 1861, inclusive. By Q. A. Gillmore, Brig.-General U. S. Volunteers, and Major U. S. Corps of Engineers. With numerous illustrations. One volume, octavo. Cloth. $4.

ROGERS (H. D.) Geology of Pennsylvania. A complete Scientific Treatise on the Coal Formations. By Henry D. Rogers, Geologist. 3 vols., 4to., plates and maps. Boards. $30.00.

BURGH (N. P.) Modern Marine Engineering, applied to Paddle and Screw Propulsion. Consisting of 36 colored plates, 259 Practical Woodcut Illustrations, and 403 pages of Descriptive Matter, the whole being an exposition of the present practice of the following firms: Messrs. J. Penn & Sons; Messrs. Maudslay, Sons, & Field; Messrs. James Watt & Co.; Messrs. J. & G. Rennie; Messrs. R. Napier & Sons; Messrs. J. & W. Dudgeon; Messrs. Ravenhill & Hodgson; Messrs. Humphreys & Tenant; Mr. J. T. Spencer, and Messrs. Forrester & Co. By N. P. Burgh, Engineer. In one thick vol., 4to. Cloth. $30.00. Half morocco. $35.00.

KING. Lessons and Practical Notes on Steam, the Steam-Engine, Propellers, &c., &c., for Young Marine Engineers, Students, and others. By the late W. R. King, U. S. N. Revised by Chief-Engineer J. W. King, U. S. Navy. Ninth edition, enlarged. 8vo. Cloth. $2.

WARD. Steam for the Million. A Popular Treatise on Steam and its Application to the Useful Arts, especially to Navigation. By J. H. Ward, Commander U. S. Navy. New and revised edition. 1 vol., 8vo. Cloth. $1.

WALKER. Screw Propulsion. Notes on Screw Propulsion, its Rise and History. By Capt. W. H. Walker, U. S. Navy. 1 vol., 8vo. Cloth. 75 cents.

THE STEAM-ENGINE INDICATOR, and the Improved Manometer Steam and Vacuum Gauges: Their Utility and Application. By Paul Stillman. New edition. 1 vol., 12mo. Flexible cloth. $1.

ISHERWOOD. Engineering Precedents for Steam Machinery. Arranged in the most practical and useful manner for Engineers. By B. F. Isherwood, Civil Engineer U. S. Navy. With illustrations Two volumes in one. 8vo. Cloth. $2.50.

POOK'S METHOD OF COMPARING THE LINES AND DRAUGHTING VESSELS PROPELLED BY SAIL OR STEAM, including a Chapter on Laying off on the Mould-Loft Floor. By SAMUEL M. POOK, Naval Constructor. 1 vol., 8vo. With illustrations. Cloth. $5.

SWEET (S. H.) Special Report on Coal; showing its Distribution, Classification and Cost delivered over different routes to various points in the State of New York, and the principal cities on the Atlantic Coast. By S. H. SWEET. With maps. 1 vol., 8vo. Cloth. $3.

ALEXANDER (J. H.) Universal Dictionary of Weights and Measures, Ancient and Modern, reduced to the standards of the United States of America. By J. H. ALEXANDER. New edition. 1 vol., 8vo. Cloth. $3.50.

"As a standard work of reference this book should be in every library; it is one which we have long wanted, and it will save us much trouble and research."—*Scientific American.*

CRAIG (B. F.) Weights and Measures. An Account of the Decimal System, with Tables of Conversion for Commercial and Scientific Uses. By B. F. CRAIG, M. D. 1 vol., square 32mo. Limp cloth. 50 cents.

"The most lucid, accurate, and useful of all the hand-books on this subject that we have yet seen. It gives forty-seven tables of comparison between the English and French denominations of length, area, capacity, weight, and the centigrade and Fahrenheit thermometers, with clear instructions how to use them; and to this practical portion, which helps to make the transition as easy as possible, is prefixed a scientific explanation of the errors in the metric system, and how they may be corrected in the laboratory."—*Nation.*

BAUERMAN. Treatise on the Metallurgy of Iron, containing outlines of the History of Iron manufacture, methods of Assay, and analysis of Iron Ores, processes of manufacture of Iron and Steel, etc., etc. By H. BAUERMAN. First American edition. Revised and enlarged, with an appendix on the Martin Process for making Steel, from the report of Abram S. Hewitt. Illustrated with numerous wood engravings. 12mo. Cloth. $2.50.

"This is an important addition to the stock of technical works published in this country. It embodies the latest facts, discoveries, and processes connected with the manufacture of iron and steel, and should be in the hands of every person interested in the subject, as well as in all technical and scientific libraries."—*Scientific American.*

HARRISON. Mechanic's Tool Book, with practical rules and suggestions, for the use of Machinists, Iron Workers, and others. By W. B. HARRISON, associate editor of the "American Artisan." Illustrated with 44 engravings. 12mo. Cloth. $2.50.

"This work is specially adapted to meet the wants of Machinists and workers in iron generally. It is made up of the work-day experience of an intelligent and ingenious mechanic, who had the faculty of adapting tools to various purposes. The practicability of his plans and suggestions are made apparent even to the unpractised eye by a series of well-executed wood engravings."—*Philadelphia Inquirer.*

PLYMPTON. The Blow-Pipe: A System of Instruction in its practical use, being a graduated course of Analysis for the use of students, and all those engaged in the Examination of Metallic Combinations. Second edition, with an appendix and a copious index. By GEORGE W. PLYMPTON, of the Polytechnic Institute, Brooklyn. 12mo. Cloth. $2.

"This manual probably has no superior in the English language as a text-book for beginners, or as a guide to the student working without a teacher. To the latter many illustrations of the utensils and apparatus required in using the blow-pipe, as well as the fully illustrated description of the blow-pipe flame, will be especially serviceable."—*New York Teacher.*

NUGENT. Treatise on Optics: or, Light and Sight, theoretically and practically treated; with the application to Fine Art and Industrial Pursuits. By E. NUGENT. With one hundred and three illustrations. 12mo. Cloth. $2.

"This book is of a practical rather than a theoretical kind, and is designed to afford accurate and complete information to all interested in applications of the science."—*Round Table.*

SILVERSMITH (Julius). A Practical Hand-Book for Miners, Metallurgists, and Assayers, comprising the most recent improvements in the disintegration, amalgamation, smelting, and parting of the Precious Ores, with a Comprehensive Digest of the Mining Laws. Greatly augmented, revised, and corrected. By JULIUS SILVERSMITH. Fourth edition. Profusely illustrated. 1 vol., 12mo. Cloth. $3.

CLOUGH. The Contractors' Manual and Builders' Price-Book. By A. B. CLOUGH, Architect. 1 vol., 18mo. Cloth. 75 cents.

BRÜNNOW. Spherical Astronomy. By F. BRÜNNOW, Ph. Dr. Translated by the Author from the Second German edition. 1 vol., 8vo. Cloth. $6.50.

CHAUVENET (Prof. Wm.) New method of Correcting Lunar Distances, and Improved Method of Finding the Error and Rate of a Chronometer, by equal altitudes. By WM. CHAUVENET, LL.D. 1 vol., 8vo. Cloth. $2.

A SYNOPSIS OF BRITISH GAS LIGHTING, comprising the essence of the "London Journal of Gas Lighting" from 1849 to 1868. Arranged and executed by JAMES R. SMEDBERG, C. E. of the San Francisco Gas Works. Issued only to subscribers. 4to. Cloth. $15.00 *In press.*

GAS WORKS OF LONDON. By ZERAH COLBURN. 12mo. Boards. 60 cents.

HEWSON. Principles and Practice of Embanking Lands from River Floods, as applied to the Levees of the Mississippi. By WILLIAM HEWSON, Civil Engineer. 1 vol., 8vo. Cloth. $2.

"This is a valuable treatise on the principles and practice of embanking lands from river floods, as applied to Levees of the Mississippi, by a highly intelligent and experienced engineer. The author says it is a first attempt to reduce to order and to rule the design, execution, and measurement of the Levees of the Mississippi. It is a most useful and needed contribution to scientific literature."—*Philadelphia Evening Journal.*

WEISBACH (Julius). Principles of the Mechanics of Machinery and Engineering. By DR. JULIUS WEISBACH, of Freiburg. Translated from the last German edition. Vol. 1. 8vo, cloth. $10.

HUNT (R. M.) Designs for the Gateways of the Southern Entrances to the Central Park. By RICHARD M. HUNT. With a description of the designs. 1 vol., 4to. Illustrated. Cloth. $5.

PEET. Manual of Inorganic Chemistry for Students. By the late DUDLEY PEET, M. D. Revised and enlarged by ISAAC LEWIS PEET, A. M. 18mo. Cloth. 75 cents.

WHITNEY (J. P.) Colorado, in the United States of America. Schedule of Ores contributed by sundry persons to the Paris Universal Exposition of 1867, with some Information about the Region and its Resources. By J. P. WHITNEY, of Boston, Commissioner from the Territory. Pamphlet. 8vo., with maps. London, 1867. 25 cents.

WHITNEY (J. P.) Silver Mining Regions of Colorado, with some account of the different processes now being introduced for working the Gold Ores of that Territory. By J. P. WHITNEY. 12mo. Paper. 25 cents.

"This is a most valuable little book, containing a vast amount of practical information about that region. It will be found useful to men of a scientific turn of mind, should they never contemplate a journey to the region of silver and gold."—*Fall River News.*

SILVER DISTRICTS OF NEVADA. 8vo., with map. Paper. 35 cents.

McCORMICK (R. C.) Arizona: Its Resources and Prospects. By Hon. R. C. McCORMICK. With map. 8vo. Paper. 25 cents.

PETERS. Notes on the Origin, Nature, Prevention, and Treatment of Asiatic Cholera. By JOHN C. PETERS, M. D. Second edition. With an appendix and map. 12mo. Cloth. $1.50.

SEYMOUR. Western Incidents connected with the Union Pacific Railroad. By SILAS SEYMOUR. 12mo. Cloth. $1.

EULOGIES IN MEMORY OF MAJ.-GEN. JAMES S. WADSWORTH AND COL. PETER A. PORTER, before the "Century Association." Tinted paper. 8vo. Paper. $1.

PALMER. Antarctic Mariners' Song. By JAMES CROXALL PALMER, U. S. N. Illustrated. Cloth, gilt, bevelled boards. $3.

"The poem is founded upon and narrates the episodes of the exploring expedition of a small sailing vessel, the 'Flying Fish,' in company with the 'Peacock,' in the South Seas, in 1838-42. The 'Flying Fish' was too small to be safe or comfortable in that Antarctic region, although we find in the poem but little of complaint or murmuring at the hardships the sailors were compelled to endure."—*Athenæum.*

FRENCH'S ETHICS. Practical Ethics. By Rev. J. W. FRENCH, D. D., Professor of Ethics, U. S. Military Academy. Prepared for the Use of Students in the Military Academy. 1 vol. 8vo. Cloth. $4.50.

AUCHINCLOSS. Application of the Slide Valve and Link Motion to Stationary, Portable, Locomotive, and Marine Engines, with new and simple methods for proportioning the parts. By WILLIAM S. AUCHINCLOSS, Civil and Mechanical Engineer. Designed as a handbook for Mechanical Engineers, Master Mechanics, Draughtsmen, and Students of Steam Engineering. All dimensions of the valve are found with the greatest ease by means of a PRINTED SCALE, and proportions of the link determined *without* the assistance of a model. Illustrated by 37 woodcuts and 21 lithographic plates, together with a copperplate engraving of the Travel Scale. 1 vol. 8vo. Cloth. $3.

HUMBER'S STRAINS IN GIRDERS. A Handy Book for the Calculation of Strains in Girders and Similar Structures, and their Strength, consisting of Formulæ and Corresponding Diagrams, with numerous details for practical application. By WILLIAM HUMBER. 1 vol. 18mo. Fully illustrated. Cloth. $2.50.

GLYNN ON THE POWER OF WATER, as applied to drive Flour Mills, and to give motion to Turbines and other Hydrostatic Engines. By JOSEPH GLYNN, F. R. S. Third edition, revised and enlarged, with numerous illustrations. 12mo. Cloth. $1.25.

HOW TO BECOME A SUCCESSFUL ENGINEER; being Hints to Youths intending to adopt the Profession. By BERNARD STUART. Fourth edition. 18mo. Cloth. 75 cents.

"Parents and guardians, with youths under their charge destined for the profession, as well as youths themselves who intend to adopt it, will do well to study and obey the plain curriculum in this little book. Its doctrine will, we hesitate not to say, if practised, tend to fill the ranks of the profession with men conscious of the heavy responsibilities placed in their charge."—*Practical Mechanic's Journal.*

TREATISE ON ORE DEPOSITS. By BERNHARD VON COTTA, Professor of Geology in the Royal School of Mines, Freidberg, Saxony. Translated from the second German edition, by FREDERICK PRIME, Jr., Mining Engineer, and revised by the author, with numerous illustrations. 1 vol. 8vo. Cloth, $4.

A TREATISE ON THE RICHARDS STEAM-ENGINE INDICATOR, with directions for its use. By CHARLES T. PORTER. Revised, with notes and large additions as developed by American Practice, with an Appendix containing useful formulæ and rules for Engineers. By F. W. BACON, M. E., member of the American Society of Civil Engineers. 12mo. Illustrated. Cloth. $1

A COMPENDIOUS MANUAL OF QUALITATIVE CHEMICAL ANALYSIS. By CHAS. W. ELIOT, and FRANK H. STORER, Profs. of Chemistry in the Massachusetts Institute of Technology. 12mo. Illustrated. Clo. $1.50.

INVESTIGATIONS OF FORMULAS, for the Strength of the Iron Parts of Steam Machinery. By J. D. VAN BUREN, Jr., C. E. 1 vol. 8vo. Illustrated. Cloth. $2.

www.ingramcontent.com/pod-product-compliance
Lightning Source LLC
LaVergne TN
LVHW050518100826
845148LV00002B/378